DOWNRIVER

DOWN

RIVER

Into the
Future of Water
in the West

*Heather
Hansman*

THE UNIVERSITY OF CHICAGO PRESS

CHICAGO AND LONDON

The University of Chicago Press, Chicago 60637
The University of Chicago Press, Ltd., London
© 2019 by Heather Hansman
Published 2019
Paperback edition 2022
Printed in the United States of America

31 30 29 28 27 26 25 24 23 22   1 2 3 4 5

ISBN-13: 978-0-226-43267-0 (cloth)
ISBN-13: 978-0-226-81997-6 (paper)
ISBN-13: 978-0-226-43270-0 (e-book)
DOI: https://doi.org/10.7208/chicago/9780226432700.001.0001

Library of Congress Cataloging-in-Publication Data
Names: Hansman, Heather, author.
Title: Downriver : into the future of water in the West / Heather Hansman.
Description: Chicago ; London : The University of Chicago Press, 2019. |
Includes bibliographical references and index.
Identifiers: LCCN 2018031113 | ISBN 9780226432670 (cloth : alk. paper) |
ISBN 9780226432700 (ebook)
Subjects: LCSH: Water conservation—West (U.S.) | Water-supply—West (U.S.)
Classification: LCC TD388.5 .H357 2019 | DDC 333.9100978—dc23
LC record available at https://lccn.loc.gov/2018031113

⊚ This paper meets the requirements of ANSI/NISO Z39.48-1992 (Permanence of Paper).

Each time we say "the River" we seem to resurrect the lost wild country.

Ellen Meloy, Raven's Exile

CONTENTS

Fish

9,080 CFS

Recreation

9,180 CFS

Future Risks

10,600 CFS

Future Plans

6,820 CFS

Confluence

3,220 CFS

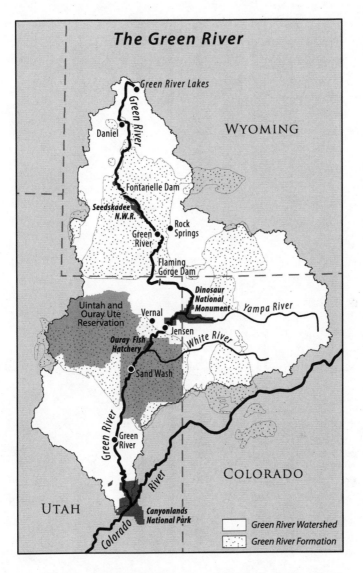

The Green River Basin. (Map created by Nicole Grohoski.)

ON THE RIVER

702 CFS

Before lunch we hit the first real rapid, Moose Creek, a narrow, jagged rock sluice that had looked fast and pinchy when we'd driven past it the night before, peering into the gorge from the road. Now, with skimpy early-season flows, there's barely enough water to split the guts of the glacially eroded granite. A few fly-fishermen flick their lines into the pools above it as we pull into an eddy upstream of them to scout the rapid, looking downriver into the gnash of whitewater. The river pillows up, then drops over a horizon line. We can't see beyond, but it looks like there's enough water to sneak our tubby little pack rafts through. Mike decides to walk around. Ben and I slide back into our boats and paddle out into center channel.

The current swells as I tee up to the top wave, pointing my bow into the break. I had been nervous, sure that I'd been off the water for too long, uncomfortable in a new boat, uncertain how my body would hold up, or whether it was stupid and stubborn to try to take a trip this long. My paddle clashes with a rock, shaking me off my line, and then I'm in the trough of the wave train, fighting to stay upright. There's a specific joy in reading water, and in knowing the micro adjustments necessary to find a tongue of fast current and thread through the shallow, ribby rapids, and my muscle memory comes back in the quick twitch of whitewater. I run through clean. It's over in a flurry of frantic paddle strokes and a smack of just-melted glacier to the face. Ben cleans it, too, and catches up to me in the slack water below the rapid. We tamp down our nervous laughs, catch our breath, turn and point our boats downstream.

We are paddling through the headwaters of the Green River, the largest, most remote, and least developed tributary of the Colorado River, which brings water to nearly 40 million people across the western United States.[1] It's crucial, it's overused, and it's at risk. There are deeply divisive battles over the water we're currently floating. The Green starts here, in the glaciated high alpine of Wyoming's Wind River Range, then winds through hundreds of miles of sagebrush flats, scrubby plains, tight gorges, and empty gas lands to the red rock desert of Utah's Canyonlands National Park, where it meets up with the main stem of the Colorado.

Because the landscape between here and there is remote and rugged, the Green and its tributaries have been touched far less than most other rivers in the West, which are overtapped thanks to human population growth and decades-long drought. It's one of the only parts of the river system with significant water left to squeeze out. But all the signs of western development—from coal to cattle to cities—are here. And as population swells in cities like Denver and Salt Lake, and as climate change shrinks stream flow, the question of how the water in this river is used and how far it can be stretched is becoming more urgent.

It's a question of whether our current way of living is sustainable in a drier, increasingly crowded West, and how it will have to change if it isn't. The Green is at the heart of that question, and I want to know what the answers might be. I'm here, white-knuckling my paddle in a stretch of water I've never seen before, because I decided the best way to understand it was the best way I know to understand a lot of things—from the river.

We launch cold and early. The sun's light cracks the ridge of the Wind River Range, gilding the edges of Square Top Mountain and glancing down onto the Green River Lakes. Mike, my journalism school roommate, and his coworker Ben, who have come along for a few days, make breakfast. We pound coffee and eat a couple of sausages each, then slide our pack rafts into the chill water on the lip of the lowest lake's beach. A Wyoming Fish and Game warden in a red shirt and a white truck, with a black dog loping alongside him, stops us on the shore. "How far you going?" he asks, as we settle into our boats and shake blood into our fingers. I swell up a little, sit taller and say, "All the way." If he notices my attempt at bragging, he ignores it, and he waves us off as we paddle along the edge of the gravelly shore. It's May, and the snow is just starting to run off the peaks that ring the basin, swelling the river. I can see the glitter of rocks on the bottom through the crystalline, sediment-free water. We make our way toward the end of the lake, and the pooled-up water starts to pick up speed. Moving slowly, but moving. "Are we on the river yet? Do you think we're on the river yet?" Ben keeps asking as we curve along the bank. We paddle under a spindly footbridge that crosses at the narrowest point. I can see the current pillowing up on the bridge pilings as it moves slowly downstream, and all of a sudden, I think we are.

I am going all the way. The boys are on board for the beginning—a weekend on the water before they have to go back to their jobs at the newspaper in nearby Jackson Hole—but I have 730 miles and a season to paddle the Green from here, in the headwaters, to its confluence with the Colorado. I'm a paddler, this river is important to me, but I feel confused about the future of water. I want to understand the complicated ways in which its water is used and where the conflicts lie in a system that needs to change to accommodate more people and less water.

The water in the Green, and in the rest of the Colorado River system, is overallocated, and has been since the 1920s, when its estimated flow was divided up between the seven states along the rivers. When the states split up the water, they did so based on incorrect calculations. They allocated more water than actually exists in the rivers, and they did it under the assumption of stationarity—that the amount of water would always be about the same. We're operating at a loss, and using more water than we have, because of embedded, decades-old policies and overstated ambitions. So far, large reservoirs, filled up in wetter times, and unused water rights have prevented the West from running dry, but we're creeping toward the bottom of our supply as things get drier and more crowded. It can't last much longer. At some point, the inflow and the amount we're using will have to be balanced, and right now no one quite knows how that will work out.

That imbalance sandpapers the space between adjacent water users. Ranchers grind against cities; environmentalists face off against the oil and gas industry. Endangered fish make things complicated for everyone. Those divides are long standing, but they're starting to be crossed, in part because they have to be. If we keep using water the way we currently are, soon there won't be enough to go around.

I feel tied to the future of water because rivers flow through my past. I started guiding rafts when I was eighteen, on the icy, dam-controlled rivers of northern Maine, near where I grew up. I learned how to run big rapids full of paddler-trapping undercut rocks in the ancient waterways of West Virginia. I moved to Colorado, looking for bigger adventures, as soon as I could, following a fellow raft guide to the mountains, and to the snowmelt streams that ran off the Continental Divide.

It was a fluke that I ended up on rivers in the first place. It started as a summer job; I just wanted to be outside as much as possible. I had no idea about whitewater, or the complicated, cross-country web of rivers, or how those rivers interlace the development of the past and the environmental management of the future. But I was quickly sucked into the paddling word.

At eighteen, I was an awkward kid, gawky in my body, unsure of my standing around other people, perpetually nervous. But being on the river made sense. I loved the physicality of pushing boats. Once I learned to read water, I could predict how they would toss me. It was a slice of grace and control that I'd never had before. And as a guide, I had to speak up. I was in charge of people for the first time, and it felt like tangible power; it made me confident, because I had to be, at least outwardly. Guiding tied me to other people, and it showed me how people affected rivers.

In Maine, we had enough water to paddle on only at certain times of day, because the dam above the rapids released water on a schedule to generate hydropower, which it pushed out to remote cities. When I moved West, the perpetual thrum of drought, and the question of whether we would have water at all, underscored every season. I don't know if I would have cared or noticed if I didn't have a stake, but once I did it became impossible to ignore.

When the seasonal swing of chasing water in the summer and snow in the winter—skiing often being the raft guide's other pole—tired me out, I became a journalist. You learn the power of an overarching narrative pretty quickly when you're in the back of a raft, in part because your tips depend on it. River guides are some of the best storytellers I know, captive audience or not, and I wanted to keep telling stories. Writing was another way to tap a stream of stories about the connections between people and places, even though it pulled me off the river.

There have been eyes on the Green River since white people first spread out across the U.S. This particular river was a pivot in the opening of the West. When Major John Wesley Powell surveyed the Colorado River down to Mexico, as part of the first U.S. Geological Survey mapping project in the 1860s, he started here, on the Green.[2] The history of water in the West has been well recorded, but its future is uncertain and underreported, shaped by hundred-year-old policy, nerves, and speculation.

The river is slow and lazy as it winds out of the lakes. The land doesn't drop much, so it pools up and bows out, barely skimming past the gnarled, wind-screwed brush that lines its banks. Ben, Mike, and I linger in those first flat miles, paddling just enough to keep our fingers warm. I keep spinning around to look back up at Square Top, behind us, above the lake. I'm not originally from here, so in a lot of ways I'm part of the problem of too many people and not enough water, but rivers are what brought me West, and what made me stay.

I moved to Colorado when I was 21, in the fall after I graduated from college, just as soon as paddling season dried up in New England. I was drawn to the rangy, idealized frontier I had in my mind, the idea of the West as a place where you could prove yourself, pushed by the landscape. This particular river had tipped off part of that obsession with wide-open places. When I was just learning how to figure out whitewater and punch through wave trains, I'd paddled Desolation and Gray Canyons, farther downstream on the Green, and fell in love with the push of desert rivers. Desolation is one of the deepest canyons in the country, a five-thousand-foot-deep layer cake of burnished sedimentary rocks, lined with sprawling sandy beaches. I spent five days wrestling the swirl of chocolatey class III whitewater, sleeping on those beaches under the star-bright desert sky, and decided I felt better there than I ever had anywhere else. The Green was everything I thought a western river should be: far off, achingly beautiful, seemingly wild.

Love can lead to obsession, so the Green became my model for what might happen to a river, for how people are going to protect a resource they've previously taken for granted and adapt to a constricted, water-poor future. But that remoteness was a paradox, too. I stopped guiding. I moved to a city. And the less time I spent on the river, the more abstract it felt, even when I told myself I was paying attention. I'd felt a sense of ignorance creeping in, and it creeped me out. Nothing I read seemed to tell the whole story of how water is tied to people's lives and livelihoods and what they might stand to lose. I decided I needed to get back on the river.

I decided I wanted to run the length of the Green, to see if I could understand the complexity of the way rivers are used. I wanted to mesh my point-source understanding, couched in recreation and my narrow idea of conservation, with reality to see what drought and overuse were really doing. And, by poring over river maps and reading trip reports from the commonly run sections, I decided I could probably paddle the whole thing and see, from bank level, the way the river was used along its course.

From the start, it feels different than it had in my head. It's not as untouched as I'd thought it would be, even in remote canyons like those I'd paddled, or here in the headwaters, where we've seen only a few fishermen. Like nearly all water sources in the western U.S., it's been dammed up, spread thin, and abused. Drought has wracked most of the western half of the U.S. since the beginning of the twenty-first century, draining reservoirs and depleting aquifers. Water is mired in the future of the climate, it's tied to the physical, political, and economic divide between urban areas and rural ones, and it's crucial to the debate about future energy sources. And more than anything, it's indispensable. More than oil. More than food.

Water feels like the biggest unstated marker of inequality and the largest looming environmental crisis. But I know that it can also be boring, slippery, and confusing. Contentious, but contained in an abstract world of infrastructure and wonky water law. Important, but easy to ignore until disaster strikes, especially for people like me who get their water piped in through city systems far removed from the source.

I'd started planning the trip the fall before, gauging how long it might take me to paddle the neglected sections that few people see, analyzing river maps and backcountry meal planning blogs, sending out frantic emails asking friends to come. Mike and Ben volunteered to paddle with me for the first few days, so I didn't have to start out alone, but there are big lonely stretches ahead of me.

My one-person-sized inflatable pack raft, which looks like a kid's puffy toy boat, is a deep, primary-colored blue, snub-nosed in the front, tapered in the stern. When I sit in it, my butt touches the back tube and my feet scrape the front. It's new to me, and when it showed up on my front porch, the box it came in hadn't seemed big enough to hold a rolling pin, much less a boat. But I'd decided it was the best tool for a one-woman, semi-solo trip: light enough that I could carry it myself, just big enough to schlep all my gear. I made a lot of decisions like that, trying to weigh potential negatives, working with minimal information. After I committed to the idea of the trip, I spent the winter thinking through risk and loneliness, keyed up on the idea of an adventure but apprehensive about how I'd handle myself if anything went wrong. I worried that I hadn't planned enough, or trained enough, or that a self-contained, source-down trip in a new boat, on a river I'd seen only a sliver of, was a stupid idea. It felt risky, but not going felt risky in a different way, because it felt like I was losing perspective. Now, paddle in hand, watching for riffles, I'm still nervous. It had all seemed abstract until now.

The summer before I set off, I'd paddled the Yampa River, the Green's biggest tributary and the last free-flowing stem of the Colorado. Being back in a river canyon, I was gut punched by how removed I'd become from that world. Even when I spent all day researching and reading about rivers running dry, and about the way the extended drought was cutting ranchers and raft guides off from their ability to make a living, I felt out of touch. I still struggled to understand the ways water rights changed hands, who was using the most water, and how global warming was bearing out on the river ecosystem. Even though I had a stake, it felt easy to skim headlines and skip the complicated parts.

I felt antsy and disconnected, and the creep of apathy started to felt scarier than drowning alone in some Utah canyon. Or at least it did abstractly from my desk. A river trip isn't necessarily a cure for indifference and self-centeredness, but I figured, in my case, it couldn't hurt. And more than that, I missed being on the river. So I started planning, trying to find a story, trying to see the stream.

All the demands on the river—the drought, the thirsty downriver cities that were buying up water and wanted more, the fish, the parks, the deepening divide between the rural economy and the urban one, the dams and the drilling rigs, the strident environmental activists, and the salty raft guides—are tied to the ebb and flow of the river, miscalculated at the outset, cut by a shrinking stream flow, and pulled thinner by a growing population. I wanted to understand vulnerability and risk, my own and other people's, and the way that water weaves through all of it.

Places like Desolation Canyon see plenty of boaters, enough that float permits are monitored and hard to get, but there are also stretches that almost no one sees, which wind through scabby, chewed-up oil and gas lands. It's not a river that gets paddled top to bottom often. It's big and weird and lonely. I'm one of the few women to run the whole thing through to its confluence with the Colorado—although I'm planning to skip the slack water of the two major reservoirs, Fontenelle and Flaming Gorge—and I'm going to do a lot of it alone. Even with Mike and Ben beside me on day one, I have an underlying thrum of long-term loneliness.

I'm scared of what I don't know, and of what I haven't seen. I'm not really sure how much I can trust my gut, or how much I may have overstated, even just to myself, how well I understand rivers. That feels desperately clear as I try to settle into my boat in the icy, early-season flow of the river. The ribs of the Winds stretch up above me, and the valley drops out below. I am in a blip of a boat that I'm worried might be leaking, swirling in a wind-rippled flush of snowmelt, hoping my calluses will come back to me, starting to paddle downstream.

FARMS

686 CFS

THE LAW OF THE RIVER

The end of the irrigation ditch is a corrugated pipe partially pinched shut by a piece of sheet metal. Randy Bolgiano, who runs the Circle Nine Cattle Ranch outside of Boulder, Wyoming, is standing in a hayfield just past the end of the pipe with a shovel, slicing off a piece of sod to block the trickle that flows out of it. The water runs into a channel along the edge of the meadow, and by blocking and pooling the flow, then pushing it out toward the field, Randy can inundate his hayfields, irrigating them. Randy and I had four-wheeled out to this end of the ditch, which isn't far from the house where he lives with his wife, Twila. It's threatening rain, the sky hanging gray and heavy, as we walk down the channel to move what he calls trash—hay and other downfall—to temporarily dam up his channels. He calls it shovel irrigating and says it's the most low-tech way to irrigate, but it seems to work pretty well; the fields are green and glowing. When I ask him why they do it this way he rolls his eyes a little and says it's because it's what they have.

There are cows with brand-new calves in the fields beyond us. Circle Nine, a cattle ranch where the only thing they grow is hay to feed the animals, is typical of the agriculture at the foot of the Wind River Range. That, Randy says, is because nothing else besides grass, not even grain, will grow out here in the high desert steppe of the upper Green River Basin. He calls the ranch a sand pile, a gravel pit, and says that flood irrigation is the only way to get a viable harvest in the narrow window between spring thaw and fall frost. The ranch has been around for fifty years, and it has rights to water through the Boulder Irrigation District, which allocates water from Boulder Creek, one of the Green's first tributaries. Seniority and source are important in the competition for water. The district has water rights dating to 1919, and Randy gets two cubic feet per second for every seventy acres of land he ranches.

Water in Wyoming, in the West, and specifically in the seven states that drain the Colorado River Basin, is monitored like this—down to the drop. Randy knows exactly how much water he gets, and where it comes from. When he wants it, he calls the ditch rider, who operates the district's canals, and asks for his water to be turned on. It flows down a series of increasingly narrow channels and eventually gets here, to the end of this pipe.

Randy is both a rancher and a member of the Upper Colorado River Commission, which oversees interstate water planning. Agriculture uses the vast majority of the Green's water, and before I get too far downstream, I want to understand the complex, interlocking legal framework that determines the way water is used across the West. Everything else about how water is managed cascades out from that. And Randy, who is tapped into the interstate legal mesh of water rights, but still needs to make a working ranch work, sees the big picture and the minutiae.

So now I'm here, four-wheeling across a flooded field in a pair of borrowed waders, trying to grasp who gets what water and why, and trying to hold onto the pitchfork Randy placed across my handlebars as we throttle out into the rain.

Across most of the West, water law, and who has rights to how much, is based on the Doctrine of Prior Appropriations,[1] which states that whoever was first able to put water to beneficial use—regardless of where they were on the river system—was entitled to primary rights. First in time, first in right, the saying goes. The doctrine stems from the California gold rush of the 1840s, when it was set up to make sure that a miner couldn't come in upstream of an existing claim, divert the water, and dry up the downstream stake. The miners knew water was a limited resource, so they instituted a standard for how western water would be allocated and when it could be used. Even if you're physically at the top of the stream, you don't get your water until after everyone who has rights senior to yours gets theirs, and if you don't put your water to use, you lose your rights to it, so you can't hoard unused water rights.

Because the Green is the biggest tributary of the Colorado River system, the amount of water available for the divvying is decided by the Colorado River Compact, a 1922 agreement that delineated how much water was in the Colorado River Basin and how it should be split up.[2] That compact is the cornerstone of a body of agreements and treaties that are referred to as the Law of the River. It sets out how much water each state is entitled to and how it's delivered to them.

In 1922, delegates from the seven states in the Colorado River Basin—Wyoming, Colorado, Utah, Arizona, Nevada, California, and New Mexico—came together because fights were starting to break out among them about where water could be diverted and dammed up. Growth in California was making all the upstream states nervous. Based on historical flow records, they decided there was approximately 18 million acre-feet of water in the river. An acre-foot is just what it sounds like: the amount of water needed to cover an acre of land—or approximately a football field—a foot deep, which is 325,851 gallons. The delegates split the river drainage into two parts, the Upper Basin and the Lower Basin, at Lees Ferry, Arizona—just below the Utah border and downstream from where Glen Canyon Dam would be built, forty years later. According to the compact, each basin was allotted 7.5 million acre-feet a year. The Green is in the Upper Basin, which includes Colorado, New Mexico, Utah, Wyoming, and a sliver of Arizona. In 1948, the Upper Basin states set up the Upper Colorado River Basin Compact,[3] which further divvied up the basin's allocation among the states. Colorado gets 51.75 percent of that water; New Mexico, 11.25 percent; Utah, 23 percent; Wyoming, 14 percent; and then Arizona gets 50,000 acre-feet. The Lower Basin is California, Nevada, and the rest of Arizona. There's no compact that governs the Lower Basin, but in the 1960s, the U.S. Supreme Court ruled that in accordance with the Boulder Canyon Act of 1928, California was entitled to 4,400,000 acre-feet, Nevada, 300,000 acre-feet, and Arizona, 2,800,000 acre-feet. The Lower Basin was also given an extra million acre-feet to expand into. In the Water Treaty of 1944, Mexico was allotted 1.5 million acre-feet a year. The Upper Basin has to make sure that 7.5 million acre-feet of water each year, or 75 million over 10 years, makes it downstream to Lees Ferry. Those hard numbers are important because the states are legally bound to them. Water rights within the states, like Randy's, are measured out based on those numbers.

It's a rigid framework for a system that's inherently variable, but that Law of the River became the foundation for growth in the arid West, where stream flow is seasonal, inconsistent, and not equitably spread out across the land. Everything that happens on the river today is predicated on the compact and how it divides up the water—there wouldn't be major cities or widespread agriculture without it. But it has one huge fundamental flaw: it was agreed upon during the wettest period in recorded history.[4] The flow was split up based on that 18 million acre-feet per year figure, but the longer historical record shows that, on average, closer to 13 million acre-feet flows through the basin each year. People who depend on the Colorado River Basin are currently using about 15 million, because that's how much the compact allocates. Even a washed-up raft guide who doesn't like math can see that those numbers don't line up. There's more water allocated than actually exists in the water system, and if everyone used the share they're legally entitled to, there wouldn't be enough to go around. Water experts call this imbalance a structural deficit. At some point, the hard numbers of the compact won't be attainable.

Not every acre-foot is allocated yet—especially in less populated places like western Wyoming—and large reservoirs have been built to store water in wet years, which is why we can currently use more than we get each year. But that can't last forever, because consumptive use is going up and nearly all the water is allocated. We're draining down the reservoirs. Both Lake Mead and Lake Powell, the major reservoirs in the river system, have been approaching their lower limits of viability in recent years, thanks to long-term drought. That's compounded by factors like climate change and the inherent variability of snowfall and stream flow.

"It's just like balancing a checkbook," says Eric Kuhn, the general manager of the Colorado River Water Conservation District, who is one of the foremost experts on the past and the potential future of the river, and one of the people I go to for help putting water use in context, thanks to a chance meeting on the river. "Over the last sixteen years, we've overused the system by 30 to 32 million acre-feet, which we know because we've drawn down storage by that amount. We started with 50 million in the bank, now we have about 18. The system is heading for zero."

Kuhn says it's both a math problem and a cultural one. Because of the use-it-or-lose-it nature of the Law of the River, which made sense when settlers were trying to develop the West, if you have water rights, there's no real incentive to use less water than you're allowed. Everyone wants to use theirs to the last drop so that they don't get screwed in the future. There's not enough water to account for that, and there's no incentive, aside from a moral one, to conserve water. "If everyone cut back a little bit you could find that space," Kuhn says. "That's the best way to do it, but it's not the way our system was designed."

Along the river, water users like Randy are trying to figure out how to sustain themselves within that system and make sure they get their water. Randy's a little bit bowlegged and a little bit shorter than me, with a sun-weathered neck and a wry sense of humor. He walks fast. I have to scamper across the rutted fields to keep up with him. He hitchhiked west from Ithaca, New York, in 1968. He's been in Wyoming since 1974, but to me he seems like part of the landscape, dressed in tan and green that mimics the field. He and Twila, who is beautiful and smiling, with shockingly white teeth, work the ranch themselves, opening and closing ditches, birthing calves, driving cattle up to higher range by horseback in the summer. They have a concrete, everyday sense of water use, and of how much they need to keep the ranch running.

It's been raining hard all spring—Randy says it's unprecedented—and the rivers and ditches are ripping past their banks. He turned his irrigation district water on last week, later than he might normally, because of all the rain. He says he'd rather irrigate with rain; it's richer in nitrogen, better for the grass, and easier, but it's also totally unpredictable, and a sustainable ranching operation depends on consistency.

That unpredictability feels familiar as I start to inch my way down-river, because I'd tried to come up with a concrete plan within a fluid system. Before I put on, I'd made a schedule, lining out when I was going to be where, working backward from a few hard permit dates downstream. But most of it—especially here, in the headwaters where few people float—was guessing. This time of year, depending on snow-melt and weather, river levels can vary by thousands of cubic feet per second (cfs), the standard measurement of stream flow. I always pic-ture one cubic foot of water as a basketball moving down the river, so a hundred or a thousand cfs can make an enormous difference in what the river looks like. In the winter I'd sat in front of my computer plan-ning a schedule that I knew could easily be washed away, trying to imagine what my days would be like and how far I could paddle an un-familiar boat down an unknown river. Now, the scariest unknowns are the environmental variables I can't control. The high peaks still have a skin of snow, which means there's still a lot to melt off.

Now that I'm on the river, my plan still feels tenuous. I still have no idea what the water levels are going to be. I'm guessing, guided only by Google, hubris, and a few kind federal land management agency em-ployees willing to answer my phone calls about how long it might take me and where I can legally camp. Luckily, so far, the river has been high. It's pushing me downstream right on schedule, even if I lollygag or paddle slowly when my shoulders get sore. But I have no idea if I can depend on that, or how much flows will change over the months it will take me to get to the confluence.

I've detoured off the river to visit the ranch, which is right on the edge of the East Fork River, another tributary of the Green. The East Fork winds along the working corral, where I can see horses grazing hay, but Randy can't use the water that runs right by the ranch. Use of the water that flows through your property is called riparian rights, and it's the standard for water allocation in the wetter eastern U.S. and across most of Europe, but it doesn't fly here. Instead, because the water is broken up by the Law of the River, Randy uses the water from the irrigation district ditch, which flows out of Boulder Lake through a series of increasingly skinny canals until it spills out of the pipe on the edge of his fields. That might seem inefficient, but that's how western water rights work.

That territorialism is baked into the Doctrine of Prior Appropriations, which also stipulates that "if you don't use it, you lose it." If a right holder is no longer putting their water to beneficial use—a use that's considered appropriate by the state water board—they can lose their right. That sets up an undercurrent of competition across the river system and between states, basins, and neighbors.

That's been the story of water in the West so far: fiercely protected and fought over.

The 14 percent of the water in the Upper Basin that Wyoming is allocated through the compact all comes from the Green. Wyoming's population is small, around six hundred thousand, and its growth is lower than the national average. The state has currently developed only about 8 percent of that Upper Basin water, slightly more than half of what it's allowed. But in planning for the future, the state engineer's office is trying to find ways to lock down all of their rights. Because of the structural deficit, they're worried that more crowded places will use up all the available water before they can use theirs, even if they still have unused rights. There are ten proposed water storage projects in the governor's most recent water plan, which was released in early 2015.[5] "Equity in the basins goes beyond the rights we have to the rights we think we have in the future. If you don't use it, who's to say the next person won't, and how is that doing any good?" Eric Kuhn says.

Because of that constant pressure to guard water rights, there's palpable tension between the more rural Upper Basin and the growing, water-poor cities of the Lower Basin, like Los Angeles and Phoenix. The Lower Basin is basically maxed out, populations in both basins are growing, and the Upper Basin is always worried about downstream users making what's known as a "call on the river," or exerting a claim for all their water before junior rights holders get any of theirs.

When I flip between country stations and NPR as I drive across the arid, sagey high desert of western Wyoming, I hear the prices of beef and oil futures on the radio. The economy is tied to what the land will give up, which is why water is so important here. Agriculture uses almost 90 percent of the water in Wyoming, and around 80 percent elsewhere,[6] but as the West shifts away from ranching toward industry and cities,[7] it's only a small slice of the economic pie.[8] It's a huge water use, and it's the most traditional one, but it doesn't add as much value to the state's bottom line anymore. If you were just to crunch the numbers, it wouldn't make a lot of sense to allocate that much water to ranchers and farmers, but those users have the oldest, most senior water rights. And the numbers don't account for feeding people, ranchers' and farmers' livelihoods, ecosystem services, or history. The current economy didn't play into the broad, legally binding web of the river compact.

Randy says that to understand agriculture I need to see where the water comes from, so we drive up to the beginning of the ditch, which takes almost an hour. He locks the hubs on an old red flatbed F250, and we follow the grid of canals out to the main road. We hit Route 191 and head north, toward Pinedale, the neighboring oil and gas boomtown that's currently busting. The Pinedale Anticline, the sixth largest gas field in the country,[9] is another significant water user in the area, but right now there aren't many new wells being drilled because prices are low. Hail plinks the windshield as we turn onto a dirt road that transects the neighboring ranches. I start to pay attention to where the fields are green and where the water pools up alongside them. Now I realize that everything that isn't sage or scratchy shrub is green for a reason. Randy points to the slight rises in the fields. "Every one of those is a ditch," he says. Once I start to notice, I can't not see them. It's an incredibly complex, ever-present network. To push water downhill, the ditches have to follow the subtle contours of the terrain. They form a latticework of potential, waiting to be filled with moving water. There's no power, only gravity. "It seems like such a huge engineering project," I say, staring out the window at the grid of green, thinking about how much work homesteaders must have put in to make their presence tenable. Randy looks at me side-eyed, like I'm slow on the uptake. "Why do you think they call it the state engineer's office?" he asks. Nothing is here by accident, it's all a carefully built, additive set of laws and landscape changes.

We end up in the granitic moonscape around Boulder Lake, the man-made reservoir that releases Boulder Creek. The hail has subsided. The fog highlights the gold in the sagebrush, and we walk through the mud across the Boulder Lake Dam. This is where Randy and Twila bring their cattle in the summer, to higher range on U.S. Forest Service land, and it's also where all the water in the district comes from.

In parsing out how much water he uses, Randy thinks about the grass he's growing and how to keep his cattle healthy, but he also thinks about the rest of the ecosystem: the migratory birds, the pronghorn, the way the saturated soil releases water through the fall and winter. His livelihood depends on keeping the land healthy, and he's looking at it every day. He says his irrigation recharges the aquifer. It provides habitat for sandhill cranes. It transforms the granite gravel pit into what he calls an oasis. Irrigation is how the landscape became livable—the present reflects the goals of the past. That's what he considers conservation.

His vision for the best way to manage water into the future is based on his reality. He hears people downstream talk about taking water out of agriculture to send to cities, and he thinks it's shortsighted, that they should focus on the top of the basin and take advantage of the natural reservoir that aquifers provide, especially when runoff from flood irrigation trickles down into the water table. "It's a hard concept in a culture that reveres conservation in the form of using less," he says. "There's lots of hyperventilation to conserve water just to comply, but it doesn't serve the entire system."

Randy isn't wrong, and neither are people like Eric Kuhn, who look at the balance of water in the basin from a more pulled-back perspective, and who think cutbacks should come from agriculture. The challenge is that it's impossible to sustain both the future of small-scale agriculture up here and large-scale water use across the basin. It's even harder when you're trying to manage for an unknown future and basing your stake on an unreliable past. The overarching Law of the River, with its built-in incentives to use a share till it's gone, makes change unappealing and risky for ranchers like Randy. It delineates how every drop of water along the river is used, but it's a broken system.

We ride back down, retracing the canals, the red truck slipping in the mud. Randy points out hardscrabble ranches carved out of tricky terrain, where getting water down the ditch is much harder than it is on his flat, even fields, and the ticky-tacky new subdivisions that sprang up to house the oil-field workers, when they were here. "People need to live aligned with the desert they live in," he says. "It's a philosophical question. Are we doing the best good by flood irrigating? I think we are."

GROWING A CROP OF
HUMANS IN THE DESERT

After meandering downriver through the craggy Forest Service land below the lakes, Mike, Ben, and I paddle into private ranchland on a Sunday. I am starting to find the rhythm of being on the river every day, the ache in my forearms has faded, and I'm less frustrated by fighting the upstream wind that seems to spring up every afternoon.

The guys have to go back to work on Monday, so in the morning we decide to leave a shuttle car at the bottom of the section we want to run. A rancher named Albert Sommers lets paddlers use a public boat ramp on his land, and we'd planned to take out there, but when we drive in to set the shuttle, the gate is closed for calving. Foiled, with part of our day cut off, we head upstream to the Huston boat ramp, which wraps around a warm, sandy bend in the river above the Sommers ranch. We leave one car there and drive the other upstream toward the one-block town of Daniel, where we blow up our boats, sling them over our shoulders, and hike down to the river.

Albert calls Randy a long cricker and himself a short cricker. The Green cuts right through his ranch, two thousand acres of rolling pastures, canal-crossed and dotted with wetlands, which slope off a bench beside the Pinedale Anticline. The high mesa of the oil field splits the landscape between Randy's place and his, a rise between the run of the river's forks. Albert's rights stem directly from the river. Instead of calling up the irrigation district, he can just open a head gate and let water gush into his ditches. A hundred years ago, his ancestors carved out these canals, gauging the tilt of the fields, cobbling together multiple homesteads until they had enough land to make a living. Now, Albert and his sister Jonita run the ranch. They've put part of the land into a conservation easement—an agreement with a local land trust stating that their water has to be used for agriculture in perpetuity.

I'd gone to visit Albert before I got on the river, and he took me out to show me what his access to the river is like. We spun out across soggy fields in a buggy, a souped-up hybrid of a golf cart and a four-wheeler that he uses to get around the ranch. He drives it fast, skidding through puddles on the open fields as we rattle out the rutted road and over the river. He makes me hop out to open gates as we cross through different pastures.

Albert, who has a gravelly, garrulous voice, wears a dusty Carhartt jacket, knee-high boots, and extra-large, metal-rimmed glasses. He has graying teeth, and when he takes off his hat, the upper half of his face is a different, paler color than the lower part. He's a lifelong rancher, and he's also been a Republican state representative since 2012. Mike, who is the environmental reporter at the newspaper in Jackson Hole—heart of the one Democratic county in an otherwise red state—says Albert is known for being thoughtful and centrist. And extremely no-bullshit. As we bump over the fields, he quickly cuts off all my stupid city-kid questions and sets me straight about the history of water use. He tells me that I shouldn't ask ranchers how many animals they own, because it's basically like asking how much money they have. He also bucks my assumptions that agricultural users steadfastly hew to the "use it or lose it" philosophy of water use, and that they ignore climate change or science. He's says he's worried about the Colorado River Compact breaking down, and about the glaciers melting in the Winds, and that he has a hundred years of history to point him toward what works and what's changing.

The settlement of the West was scratched out on land like Albert's. In 1862, President Lincoln signed the Homestead Act,[1] which guaranteed settlers 160 acres (a quarter section) of free public land if they had the tenacity to live on it for five years. Pioneers headed west, looking for a quarter section to claim as their own, working on the misguided assumption that as agriculture crept across the arid western plains, rain would follow the plow—a common theory at the time, put forth by land surveyor Charles Dana Wilber, who thought farmers could change the climate by changing land use.[2] That kind of acreage was enough to sustain settlers in the rain-heavy East, but west of the 100th Meridian—the north-south line that splits the United States and demarcates the desert West—160 acres was rarely enough to grow a sustainable crop to feed a family. Plus, in the tricky topography of the western ranges, farmable claims were hard to find. The chalk-dust plateaus and gravelly sagebrush hills along the Green were some of the last places to be settled. In 1869, when a one-armed Civil War veteran named John Wesley Powell set off to navigate and map the Green and Colorado Rivers, the area around them was the biggest remaining blank spot on the U.S. map. Powell made recommendations for how land and water should be used that remain spookily relevant even now. When he published his account of his trip, *Report on the Lands of the Arid Region*, in 1878, he advised western amendments to the Homestead Act, such as breaking up claims by watershed boundaries instead of acreage, that would have changed the way the West was developed.

That's because to grow anything substantial west of, say, Kansas, you can't depend on rain. You need to bring in water. Rainfall is scant and stream flow is inconsistent. Irrigation and access to water quickly became the key to survival in the West. Cities like Denver, on the South Platte River, were sited for their access to water. You can tell where the oldest homesteads are in any given valley because, like the Sommers ranch, they're at the junctions of streams. Albert and his sister have a homemade museum on the site, where they bring in school kids and tourist groups to tell them the story of how the land was settled, and what it took to make it last. "When Lincoln signed the Homestead Act it was the biggest experiment in social Darwinism in history," Albert says. "It's not possible to make a living on 160 acres here. The people who are left 150 years later are the ones who survived."

The river feels different here among the ranches that roll out from the water. The current has picked up. Side streams, like Horse Creek and Forty Rod Creek, flow in upstream, and the water is warmer and murkier. I can barely see the gravel bed of the bottom anymore.

We slide downstream without doing much work, paddling loose-shouldered and easy.

The floodplain is flatter and grassier, with more cottonwoods and less scrub. The ridgy mesa of the Pinedale Anticline tips up to our left, scrubby and rolling, layered deep underneath with frackable oil and gas. The banks are sandy clay cliffs, spackled with swallow's muddy nests. "It feels like there's a lot more wildlife here," Ben says, binoculars around his neck. We spot a huge white pelican with black-tipped wings, heavy-headed moose, and the bluest, greatest heron I've ever seen. I'd thought of herons as being native to the brackish eastern estuaries I grew up on, but I start to see them everywhere on the river. I don't have enough history to know if the throngs of wildlife are here because of the irrigation, or if this is how it's always been.

The longer I spend at eye level to the riverbanks, the more details I notice. This corner of Wyoming is full of superlatives: largest remaining sage grouse habitat, biggest gas field, least populated. We're a rim of peaks away from Yellowstone National Park, one of the most visited in the country, but the Wind River Range, which seems equally beautiful to me, is effectively empty, and is not federally protected in the same way. What counts as valuable seems arbitrary; the human demarcation line seems blurry.

Even for an expert, it's hard to tell what's natural or native at this point because the changes in the ecosystem are similarly subtle and blurred. Ranchers argue that irrigation smooths out the seasonal variation in stream flow and recharges the aquifer, creating habitat, making it better for people and wildlife. But it's not necessarily historically native animals that would be there if ranchers weren't watering the land. That's the tricky part of trying to recreate an unclear past.

They're trying to even out an ecosystem that's inherently spiky. Ranchers, who are not given to gauzy language, use the term "ephemeral" a lot when they describe streams to me—the creek is there and then it's gone, sapped dry after snowmelt. Use it while you can, and try to hold onto it because it'll be gone soon.

Like Randy, and most of the ranchers around here, Albert flood-irrigates his fields to grow grass for cattle. In the summer he moves the cattle up to the Green River Lakes by horseback to graze on the Forest Service land up there, and in the winter he brings them back here, to what he calls the home field, to eat the hay he's growing now. This type of farming is often called low-value agriculture[3] because it uses a lot of water to produce minimal results, but it's what families like the Sommerses have been doing for generations. The hayfields their ancestors seeded with timothy and Alsace clover have never been re-seeded; they're still working on history. It's the way they know how to make a living, they're confident that it maintains the health of the eco-system and the local wildlife. They feel a generational claim to those seasonal, stream flow–based cycles.

Back on the river, we skim through riffles, barely paddling. At one point, Ben and I decided to go right around an island, chasing micro-rapids. Mike, who has his dog in his boat, goes the other way. We come out after a long meander and don't spot his lime-green boat, but rivers only flow one way, so we paddle on. We float for long enough that I start to question how far we've gone, but river miles are slippery things, constantly in flux and tricky to gauge, especially on a stretch of stream you've never seen before. We paddle until I recognize a red roof in the distance. Albert's place. We realize we must have missed the takeout during our detour, and now we have two options: jump the locked gate and trespass, or float on for an unknown number of miles to the next boat ramp. It took me a while to get into Albert's good graces, to convince him that I was smart enough to understand how to get by around here. I've proved myself wrong on that front, but we decide it would be stupid to go on because we don't know how far down the next take-out is. We jump the fence and walk up the road with a desperate lack of subtlety, our bright green and blue boats catching the breeze as we chuck them over the gate. Ben scuttles through the fields in a wet suit, and I'm in a bright orange dry suit. It doesn't help that Albert spent a lot of our conversation impressing the importance of good neighbors, and respect for other people's property, on me.

In the past, water monitoring along the Upper Green has depended largely on handshake agreements. Albert says he can go talk to his neighbors if it feels like he's going to run low, and they're still a little cowboy with the measuring. Occasionally, in past dry years, that casualness has devolved into fistfights or stealthy outings in the middle of the night to open other ranchers' head gates, but for the most part they keep it civil, because everyone knows one another well. Both Randy and Albert say they know they can currently get away with using more than their share because they're so far up in the headwaters. Everyone knows how much they're supposed to have, but both regulation and monitoring are still a little loose, especially when there's enough water to go around. It's a mix of historical use, a bit of regulation, and some neighborly peer pressure. To me, dropping in from afar, where all I've been hearing about is basin shortages and drought, this informal management system seems nice, but unrealistic over the long term, especially as supplies tighten downstream.

Those handshake agreements that hold together the old ways of doing things are tenuous and aging out. According to the National Young Farmers Coalition, farmers over sixty-five outnumber those under thirty-five by a ratio of six to one, and nationwide, over 573 million acres of farmland are expected to change management in the next twenty years.[4] You can't hold onto history forever. Albert might be a part of the last generation that can eke out a living the same way their ancestors did. That's part of the reason the Sommerses put some of their land in a conservation easement. They want to preserve the way they do things into the future. "I see the end of flood irrigation and I hope to God it's not in my time," Albert says. "Laws are going to change, compacts are going to be broken, and I can guarantee you that people who live on the land, who live in this landscape, are not going to be the beneficiary of that."

Neither Albert nor Jonita had kids, so they brought in the "neighbor boy," Ty Swain, to slowly take over the ranch. After hearing about him, I'm surprised that Ty is a grown-up with a wife and baby. When I'm roaming the fields with Albert, the Swains meet us in a different buggy with the kid in a car seat covered by a blanket, and we all go to open up the head gate of a ditch that comes right off the river.

There are wide wooden boards jammed into the shoulder-high concrete casing, stopping the flow. Albert has a long, skinny metal hook with a spike on the end in hand as he and Ty each kneel on one side of the gate. They hook the top board and pull it up out of the frame of the gate, letting more water spill over the top. Albert watches the water level come up in the ditch and says, "Better pull another." He can gauge how much water he needs downstream by eyesight. That ranch is their world, they have all the patterns memorized, it's an ingrained sense of the land on a micro scale. "You can pet that one," Ty says to me, gesturing at one of the all-black cows we roll by in the buggy, "but not that one," pointing to another, seemingly identical one.

Agriculture gets vilified for being outdated, for using all the water, for waste, but flood irrigation has deep roots. It's tied up in decades of public land use and water law, and in the myth of the American West— that ranchers like these tamed the wild and made us what we are. It's hard to untangle from that history, and agriculture in the West isn't just flood irrigation for grass here in the high plains. It's almond growing in California and melons in the desert of Utah, too. In Wyoming, agriculture is the biggest water user in the state. The farmers were there first, so, according to the Law of the River, they get their rights first. And, as many aging farmers argue, we need food. How are people in cities going to eat if all the water moves out of agriculture to the cities?

Studies have shown that in the current agricultural market it still makes economic sense for ranchers up here to use flood irrigation, at least until the price of buying water outweighs the cost of buying hay elsewhere.[5] That makes it tough for people like Albert to see the value of changing their ways, even though they know there is greater need for the water. "I think the ranchers are resistant to change, even though they know it has to come, because it works for now. And they know what works, because they have to make it work every day," he says. "I'm pro water development if it's done right and people's interests are protected, but it's hard to want to change when you don't see the positive outcome."

Ben and I hit the road by Albert's mailbox without seeing anyone. We deflate our boats and figure we can hitchhike back to the boat ramp to meet Mike. But it's Sunday in rural Wyoming. After an hour, no one has come by, and Ben decides to start running for the car. He takes off jogging in his wet suit and river sandals, and I stay with the boats, sitting low in the scrub on the side of the road. Shortly after he leaves, I hear an engine thrumming up the hill. Albert pulls up in his truck. "It is you!" he says, as I bumble through an apology about how we'd missed the takeout, and how, in normal circumstances, I totally respect private property. "It's fine, you have permission," he says, underscoring the handshake agreements that make things run smoothly around here. "I talked to you, I know who you are."

I chuck the boats into the back of the truck and we drive north, stopping to pick up Ben, who is sweating in his wet suit on the side of the road. As we bump back toward the boat ramp, Albert says he's frustrated because he thinks everyone involved, from local water users to national politicians, is looking at the problem only from their own side and not working hard enough to find an approach that serves everyone. Because there are such embedded assumptions that the other side is going to screw you over, he says, it's hard to get everyone at the table.

I can picture Albert debating in session, because he's already upended my thinking about the subtly different ways people use water, and about how long it takes to build consensus. That drive to make sure every voice is heard is why Albert is a savvy, realistic politician. He has a concrete sense of all the other potential water uses and the battles on the horizon. He's aware of the golf courses and subdivisions that are likely to spring up downstream, and he knows that the glaciers are melting upstream, but he's also deeply committed to the life he knows. He wants to make sure the historical uses, which incorporate farming and fisheries, still hold weight in a changing world. He says he's trying to hold onto the life people in the valley have had, and still have, while preparing for the future. He's not necessarily stonewalling change, he's just careful about what he thinks is best. And he's coming to it with his built-in history of living on the land, dependent on the ebb and flush of the river. "I don't think that people look at all sides of the story," he says. "I don't think we as ranchers are willing to look at the fish perspective and the fish aren't looking at us. A city can't understand why they can't get water when they're growing a crop of humans in the desert."

ALL THOSE PEOPLE HAVE TO EAT

I'm on the river alone once Mike and Ben head back to work. I'd been worried about the day-to-day grind of life on the water, that I'd be bored or scared or frustrated, but once I've slipped the bonds of social appropriateness, I slide into a routine I start to love. I wear the same thing every day: Pink board shorts, quickly faded to brown by the rust of the river. A plaid snap button shirt once I get smart enough to shirk the sun. The same blue ball cap, and a braid. I have perpetual thick white schmeers of sun block on my face, but I don't figure that out until I look at pictures later.

I'd never done a trip like this before, nothing this long, or this lonely, or where I've had to make every decision by myself. At home, the details of planning the trip had felt overwhelming and unfathomable, even deciding what food to bring was fraught, so I'm surprised at how easy it seems now that I'm here. I end up eating the same thing almost every day: packets of oatmeal and coffee for breakfast; tortillas, cheese, peanut butter, and apples for lunch; Tasty Bites for dinner; Pringles and granola bars in between. I turn off the part of my brain that usually obsesses about food. When it gets hot and stays that way, I crave beer and salt and popsicles, but other than that I boil water when I wake up and don't think about much else.

But I can't stay in the suspended disbelief of river time forever, so after a few days I take a break from paddling and drive east, to the border of Wyoming and Colorado, to visit Pat O'Toole, the president of the Family Farm Alliance, so I can get a feel for how conservation and agriculture can work together in the future, and how that can make ranching sustainable. He ranches at the confluence of Battle Creek and the Little Snake River, which intertwine in a lush, peak-rimmed valley. The two streams eventually run into the Yampa River and then the Green, crossing state lines a dozen times before they do. Pat, who looks like the platonic ideal of a rancher, down to the bushy white mustache and the bandana around his neck, has been one of the most outspoken voices in the country for trying to align food production and conservation.

Ladder Ranch, which has been in Pat's wife Sharon's family for six generations, is the greenest place I've been in a long time, a contrast to the silvery sage of the Green's headwaters. Fishermen picnic at the confluence of the two streams, cattle and sheep dot the far fields, grandkids in tiny cowboy boots skitter in the barnyard, and Sharon is making tuna sandwiches for neighbors who stopped by.

Pat and I jump into a buggy similar to Albert's, and he takes me out to see his favorite spot on the ranch. We ride out to the confluence of the two rivers, where he's built a pool that acts as a small-scale water storage structure and a fish habitat. Ladder Ranch, like Albert's ranch, is in a conservation easement, and Pat has worked with The Nature Conservancy and the Wyoming Fish and Game Department to tailor his irrigation system and stream flow management to help out fish and recharge groundwater while still providing enough water for his fields.

The battle over water often pits groups against each other: farms versus fish, or farms versus cities. One needs to lose water for another to gain it. Pat says you can support all the parts of the system, the key is keeping the ecosystem in balance. He thinks that the way they're doing it in the headwaters, through small-scale storage and habitat creation, is a model for getting all the parties on board, for making sure that all the water users are considered. He doesn't think the idea of all or nothing—of taking water away from a ranch and sending it all to a city—works for anyone. His model works well here, in the headwaters, but I wonder if it misses some of the large-scale pressure that is impacting water use downstream, He doesn't have anyone upstream of him who might impact his use.

Pat says his focus is natural alliances and trying to get people to agree to work together toward common goals, and he's seen enough to have a targeted plan. He's leery of new policy, and he doesn't think more regulation is the answer. People who want to sacrifice food for urban growth or energy development trouble him. "I give speeches about the hopefuls and the hatefuls," he says. "Conservation groups that work with agriculture, those are the hopefuls. The hatefuls are the litigators who want to get water off the land. If we don't understand strategic watershed storage, we're not understanding food and we're not understanding our country."

The fear of water being pulled out for other uses comes from several places: There's the Colorado River Compact, which says that the Upper Basin must send the Lower Basin its yearly 7.5 million-acre-foot share, regardless of precipitation. And there are current economics and demographics. People are moving to cities, and, because people need to drink, water is moving with them. The hard lines of existing water law, which were designed to protect the rights holders, make it difficult to share water or spread it around. If you can ignore the ways we currently do things, incremental change and increasing flexibility are the most obvious and sensible ways to make sure multiple water users get a fair shake, but that's not how the Law of the River currently works. The complicated part is getting all the users to agree on how to be flexible and to work within the existing legal confines of the compact and state law.

The unmitigated forces of growth and drought exacerbate the problem. The Colorado River Basin has been in what the Department of the Interior calls a "historic extended drought" since 2000,[1] and a good number of the fastest-growing cities in the country, like Salt Lake City, depend on the basin, and on the Green in particular. Then there's the lynchpin that people like Pat are pivoting on: we have to eat. A recent report from the United Nations said that estimated worldwide increases in population will require a 70 percent increase in food production by 2050.[2] Pat says he worries that in planning for the future, and in pulling water out of agriculture, people forget that agriculture is related to food. And that we need food to survive.

It takes a lot of water to grow things. The USGS estimates that it takes 460 gallons of water to make a Quarter Pounder, for instance.[3] That's a sticking point in the farms versus cities fight: there are higher-value and lower-value ways to use irrigation water, and one could argue that the amount of water used to grow beef in Wyoming—which is a lot—could be better used in other parts of the river system, like in California, which grows a third of America's vegetables and two-thirds of our fruit. But in places like rural Wyoming, where people like Pat, Albert, and Randy have staked their livelihood on agriculture and have senior water rights, they can't grow much else besides grass. Changing water allocation might unravel the social fabric of those small towns.

In 2012, the U.S. Bureau of Reclamation published a study that said the gap between water supply and demand in the Colorado River Basin was anticipated to approach or exceed 3.2 million acre-feet by 2060. That's almost half the water allocated to the Upper Basin each year.[4] Because agriculture is the biggest water user in the basin, it makes sense, in terms of scale, to cut that deficit by reallocating that water to other places. "In most parts of the West, reallocating just 10 percent of agricultural water to municipal uses is generally sufficient to boost municipal supplies by 50 percent," says Doug Kenny, the director of the Western Water Policy Program at the University of Colorado, which is trying to find the most realistic, unbiased solutions for future water use. He says that incremental changes in use could actually make a big difference downstream.

Pat thinks those incremental adjustments will permanently change the way we grow and price food in the country and how we manage our ecosystems. He's worried that when cities look at agriculture as a new, untapped source of water, they're ignoring both potential environmental and economic problems. In his mind, the consequences of those adjustments would be too high to allow because they'd set off a chain of actions that would be hard to undo. Once a water right is gone, it's gone. If at some point enough agricultural water users give up their rights to cities, it could mean a lack of water to grow food, or a depletion of instream flows, which would affect fish and other parts of the ecosystem. "If you keep your farms and lose your cities, your cities will still regrow. If you lose your farms and keep your cities, you lose both of them," he says.

With Pat's perspective in mind, I head back to the river to paddle the rest of the free-flowing section of the Green alone. At this point, the river has browned up. It's loaded with the green-gray alluvium of ancient glaciers, which sloughs off the surrounding fields. It's still early in the spring runoff cycle, and the water changes every day, pulsed by upstream snowmelt. Now, it's wide and gentle. The river has been relatively flat and the paddling free of consequence so far. I'm mainly dodging rocks, watching the horizon line for any sense of drop, feeling the current pull me around corners. I camp on tiny slivers of high-water beaches, pulling my boat high up on the shore in case the water rises in the night. I scan Bureau of Land Management maps for pockets of public land, trying not to inadvertently trespass again. My next destination is Fontenelle Reservoir, formed by the first dam on the river, one of two major storage sites on the Green.

My pack raft is about as long as I am tall. I paddle it with an old, chipped kayak paddle, bought in the days before I realized I was too nervous to be any good at whitewater kayaking. The little raft feels slightly more stable than a kayak and is easier to wrangle than a canoe or a full-sized raft, but it's easily deflected by wind, and it feels flimsy sometimes when I'm alone. For the first few weeks I'm constantly worried that it's deflating on me.

When I'd planned the trip it had seemed easiest to do it by myself, so that I didn't have to hew to anyone else's schedule. That solitude felt abstract when I was looking at river maps and calculating mileage in front of my computer at home, but as my launch date grew closer, I started to go through all the bad things that could potentially happen to me: flipping and losing my boat and all my gear, hypothermia, dislocated shoulders, murder. Friends, trying to be helpful, told me stories about women alone in the wilderness stalked through their GPS tracks or followed for miles along trails. Lying in my tent alone at night, it's impossible not to think about those stories. Mike and Ben alleviated my fear for a while, but now that I'm fully alone, it creeps in again. I'm getting less jumpy every day, but there are still nights I lie awake listening to every noise, river knife just outside my sleeping bag, just in case. My worry about my physical ability to paddle has subsided a little every day, as my fingers callus and I keep making miles, but the twitch of unknown outside risk is still under my skin.

That tiny undercurrent of anxiety means I have my eyes open all the time. I'm tuned to small shifts in the current and the micro-scale changes in the scenery. The river makes a break in the landscape between bluffy eroded cliffs, striped red, white, and green, on the left bank and pivot-irrigated fields on the right. The valley feels like the long curve of a wind tunnel, scrubbed open by sun and sky. Tributaries fan in across the Upper Basin, cutting through the bedrock and the brush, and I can feel the desert coming downstream. Above the town of La Barge there are big fins of red rock alongside the river. The valley spreads out and the mountains suddenly feel far away as the topography turns to buttes and wide, cottonwood-lined river bottoms. The river weaves through one-strip towns, like Marbleton and Big Piney, still paralleling the ridge of the oil field, where drilling rigs appear in the distance. In a lot of those towns old homesteads have been subdivided into ranchettes—plots of land that are too small to ever grow a sustainable crop, and which will now no longer have to. Depending on where they are, subdivisions generally use about the same amount of water as agriculture, but it's all consumptive use, it's not percolating back down into the water table.

"The value of western working land is measured by the value of water on those lands, and water is moving out of ag daily," says Kate Greenberg, the Western Water Program director at the National Young Farmers Coalition. Kate says it's scary for famers to talk about flexibility because their livelihood is tied to practices that they've spent decades perfecting. "Farmers only have one chance every year to try out their ideas. It's not like the tech industry when you can try something all the time," she says. It's expensive to make changes, both initially, when things like pivot irrigation systems or digital head gate monitoring can be huge investments, and potentially on the back end, when a long-term lack of water can cut productivity and the bottom line.

Even in the hundred or so miles I've traveled—a fraction of the river's path—the way people use water has changed. Here, where it's a little bit warmer, flatter, and less gravelly, and where the growing season is slightly longer, ranchers can run sprinklers and use less water, unlike Randy or Albert. The craggy range has smoothed out just enough to make a difference. Agriculture is an industry built on microclimates: it can vary on two different sides of the same hill, and minuscule changes in how much water is put on the land can change things significantly. What makes sense for water use at the top of the river doesn't necessarily work farther down.

It's warmer here, and the hills are lower, so it feels like there's more light at the end of the day. As I paddle past La Barge I can start to feel the river pool up as its movement becomes slower and swirlier. I'm just upstream of the Fontenelle Dam, which backs up water for a dozen miles. It's the first real break in the river, and I can sense it upstream. Things are changing. From here, the river is human-managed, controlled in part by the release of the dam.

Everyone I talk to says they know change is coming; they're just not sure how to deal with it. They're trying to conserve. No one is willfully wasting water, an accusation that often gets thrown at farmers who use flood irrigation, but conservation means different things depending on your perspective. It doesn't necessarily mean using less water; sometimes it means preserving a habitat, or a way of life. The unknown future is scary because we're both beholden to and formed by the past. It's alive and important here; tradition and nostalgia hold weight. "No community is different, you're either growing or you're dying, there's no in between," Pat says. "Water is one of those drivers and always will be."

CITIES

2,790 CFS

THE ONLY WATERING HOLE
IN THE WHOLE COUNTY

The river hits the highway and its first real city at the same time. Below the Fontenelle Dam, it drops slowly through the Upper Green River valley, curving through the buff-colored hills of the Seedskadee National Wildlife Refuge. It takes a big bend under the semi-trucks rushing across I-80 before splitting the city of Green River, Wyoming.

Unlike farms, which get water through rights tied to agriculture's designation as a beneficial use, city dwellers rely on municipal water organizations. In Green River and Rock Springs—two neighboring cities that, combined, make up the fifth biggest metropolitan area in Wyoming—water is pulled out of the Green and processed into drinkability by the Joint Powers Water Board. That public utility is the single source of water for the two cities as well as for most of the industrial and mining use in the rest of Sweetwater County. That's because, in the drought-cracked high desert of western Wyoming, that's pretty much all there is. "We're the only watering hole in the whole county," Fred Ostler, the General Manager of the JPWB tells me as we stand outside the water treatment plant, baking in the early June sun. "This is the iconic West and there's not a lot of water around. Welcome to the land of not green."

Fred walks me through the plant, which sits beside the river on the southwest side of the city of Green River. It's a series of massive green-roofed buildings connected by a string of underground pipes. We walk to the riverbank and start where the treatment starts—where 40 cfs of river water is pulled through intake tubes that look like huge silver air ducts. When I'd paddled past a few days earlier I'd heard them whir-ring, churning up foam.

Rock Springs and Green River, which both ride the economic growth curves of the energy industry, feel run down and highway worn, a little thin around the edges. I'd spent the night camping at an RV park in Green River so I could take a shower, and the brother of the owner stared at me long enough that I got jumpy. It seemed like he didn't see strangers much. Rock Springs, the bigger of the two cities, sprang up because it was a source of coal for the railroad, which chugs right through the middle of downtown. There's no water source in the city except for an ephemeral stream called Bitter Creek, which is exactly as appealing as it sounds, so Rock Springs pipes water from the treatment plant in Green River twenty miles away. Everyone who lives there de-pends on the single pipe that transports the water from the river to the city.

From the edge of the river, Fred and I walk over the underground pipes where the water is treated with chlorine and ozone to break up pollutants. He tells me the same joke—that, like morticians, they bury their best work—twice. As the river water is piped into the plant, they add iron sulfite and polymer to make any particles coagulate, then they skim them off. From there, the water cycles through a series of mas-sive indoor pools: a flocculation basin with a mud blanket at the bot-tom, which helps with the coagulation; then filtration beds, which suck out tiny particles with granular activated carbon. Together, the pools act like a massive sponge, sucking up grime. The inside of the building has the damp tang and high ceilings of a city swimming pool. There are lifeguard rings on the aqua walls, and some of the plant's thirteen employees are sitting in the control room above. They need guard-like vigilance, because if they lose focus, or if something goes wrong, it could cut off water for fifty thousand people. They're currently work-ing on building backup systems, including a thirteen-acre off-channel reservoir, in case something pollutes the river, as well as a second pipe to get water to Rock Springs in case the first one breaks, but for now, they're the only line of defense. "Everything comes through here,"

Fred says. "We're watering our grass with drinking water, but it's the only choice we have."

Dry urban centers are facing increased pressure to get water to their residents. According to the U.S. Census Bureau, population in the western states grew by 20 percent in the 1990s, mainly in Lower Basin cities like Los Angeles and Phoenix, and that growth isn't slowing down much. As of the 2010 census, growth in urban areas is outpacing the overall growth rate, and nine of the ten most densely populated urban areas are in the West. The Southwest is the fastest growing region in the country.[1]

People in those growing cities need water, and there's a gap between where the water is and where it's needed. Most of the water west of the 100th Meridian falls as snow and rain in the high country, then drains out of the mountains and into the rivers. The Wind River Range averages between 40 and 60 inches of precipitation a year, while the rest of the Wyoming gets about 14.5 inches. The supply and demand are geographically mismatched. While the image of vast, empty western ranges might still be true, it's not because there aren't that many people in the West. It's because of a divide between increasingly sparse rural populations and increasingly dense urban ones. That divide shows up in politics and demographics, and it also shows up in how people use water.

To get to Green River I've floated a chunk of water bracketed by two dams, Fontenelle above and Flaming Gorge below. Since Mike and Ben left a few weeks ago I'd been alone in the sere desert above the Fontenelle Dam and Reservoir, but now I have company. My college friend Will had never been on a river trip before and hadn't camped in a decade, but he was feeling city-trapped and in need of some adventure, so he flew from Washington, DC, to Salt Lake City and drove across the openness of southwestern Wyoming to meet me. He brought his roommate from DC, Dutch, who'd grown up in rural Oklahoma but had shoehorned himself into a government job. Dutch wanted wildness in the same way. I meet up with them in Green River, and when they drive into town, their car is stocked with Cheetos, whiskey, and not much else, and they're hours late. They'd forgotten to rent a boat in Salt Lake City—the one thing I'd told Will he had to take care of—and ended up driving hundreds of miles out of the way to find one. They'd underestimated the emptiness of eastern Utah, and then western Wyoming, and Dutch is somehow already sunburned. After being on my own program for so long, I'm annoyed by their shitshow, impatient with their bumbling, but I lose my edge a bit when I see how excited they are. We grocery shop in town, leave one of the cars at the park where we'll take out in a few days, and then drive back up to the Fontenelle Dam, where we'll start paddling.

The first night we set up camp right below the dam. Will, shivering in the only layers he brought—a cotton T-shirt and tinselly windbreaker—runs up the edges of the high banks, trying to take in as much of the sagebrush and sky as he can. I had tried to explain the empty ranginess to him on the phone. The way that being attuned to the micro-riffles on the river feels concrete and purposeful in a way lots of things don't to me these days. It feels corny and kind of impossible to say that out loud. But now that he's here, I think he feels it too. "I didn't know places like this existed," he says as sunset scuds the low clouds to the west, turning everything gold.

Our plan is to put in right below the turbine of the Fontenelle Dam, paddle through the Seedskadee Wildlife Refuge into the city of Green River, and then into the slack water of Flaming Gorge Reservoir, right on the border of Utah. The refuge was set aside as a mitigation measure when the Fontenelle Dam was built, to ensure that some habitat remained after the dam went in and the reservoir filled. It's a random place for a wildlife refuge, I think, but it made sense, logistically, in relation to the dam when it was built. It feels less dramatic to me than the glacier-scoured valleys that flowed out of the Wind River Range—it's browner, flatter, and less beautiful to my already jaded eye. But cities need water, and that need alters the rest of the landscape.

If you look at it with unsympathetic eyes, the future of water is driven by economics. That's true from the headwaters to the confluence. When I rode around with Albert, he repeated an old adage: Water flows uphill to money. It's not just an expression. In 1922, when the states determined how much water they would each receive, they looked at the river like a plumbing system, usable down to the last drip. In developing municipal water systems, demand has more weight than hydrology or instream flows. Water was, and still is, pumped over whole mountain ranges to serve cites that can pay. That makes places like Green River worry that their water might be diverted before they can use their full share. In Utah, much of the Green's water is piped to the growing Salt Lake City area, bypassing less populated rural areas and tribal reservations. There's a deep fear that urban demand will dry out less privileged areas. Once there's not enough to go around it's hard to dewater systems that are already set up.

The pace of my river days is different with people along. We paddle through the pools and drops, trying to catch current so we don't have to work too hard. Days stretch out in an ever-expanding way. What does that cloud remind you of? How many of the words to '90s rap songs can we remember and sing? How would we look at the world differently if we'd been born here in the empty high desert? Will sees his first pronghorn, and we float past herds of wild horses. Some days we crack beers before noon and stash our paddles, moving at the speed of the current. We push ourselves out of eddies if we need to, but otherwise let the river take us. We camp in the scrubby brush along the banks, fighting mosquitoes and wind. One night, by the side of a ramshackle bridge, Dutch cooks us his dad's camping specialty, fried Spam, for dinner, and in the low light of the oncoming summer evening, we cross the bridge and poke around the ruins of an old mining cabin on the other side, trying to imagine the people who once lived there. We're not the first people to visit since then; there are bullet holes in the bridge and buildings and a litter of empty cans scattered in the foundation.

Will says that even the roughed-up abandonedness feels interesting. He's not used to the skies and buttes and emptiness, or the way you can watch the day blow through the landscape. His awe underscores a lot of the fights I've been seeing between rural and urban places, which stem from a lack of understanding of a place you've never seen. It's hard to internalize priorities until you've felt them.

The river slits the bluffs of the Red Desert, one of the largest unfenced landscapes in the lower forty-eight. It's huge and harsh. We get battered by wind and sun, blasted by sand, and thronged by bugs any time we stop. From the river we see moose, pronghorn, and more birds than I can recognize. This desert is home to sage grouse, wild horses, and the largest migrating herd of those threatened pronghorn, but it's a hard place to make a living.

Cities like Rock Springs and Green River are hollowing out, especially while oil prices are down and there isn't much work on rigs. There's a trona mine that provides some steady work, but people and jobs are leaving for bigger cities where there are more economic opportunities. When I walk through the water treatment plant with Fred, he tells me that they're currently operating at about a third of their capacity, in part because of the huge fluctuation in the number of customers. "Rock Springs went through a huge boom, but now it's back in the bust cycle," he says. "We have some companies that are very stable and have been for years, but there are gas and oil wells, and coal mines, that haven't been able to operate." The area's population is about fifty thousand, more than 10 percent of Wyoming's residents, but it's a small pocket compared with a place like Salt Lake City, where more than a million people in the metro area depend on the same river sources. Fred says Sweetwater County is in good shape, supply-wise, for now, and that its water rights are some of the older ones in the state. But, like most water users, its residents and water managers are constantly thinking about the future. They're nervous about shortages in the upstream supply, and about proposed transbasin diversions, which would send water out of the river to big, growing cities like Salt Lake and Denver. Transbasin diversion is the umbrella term for water projects that take water out of one river basin for use in another, where there might be a bigger need, or a greater financial gain. Such diversions fight gravity and hydrology, and the larger ones are heavily engineered—they often pump water through range-wide tunnels or up over mountain passes. Urban growth in the West would have looked very different without them. In a lot of western cities—many of which, like Rock Springs, were founded based on their mining potential, not necessarily their liquid assets—these diversions are a way of guaranteeing water for their residents. They're complicated for a lot of reasons, not the least of which is that they can significantly change water access downstream.

The supply chain is a concern for smaller cities, too. The area around the city of Green River gets about six inches of rain a year and has no groundwater, so residents are trying to guarantee that they'll have their water rights if or when the city starts booming again, even if downstream cities grow faster than they do. "We're trying to say, don't take our water, or if you do, pay us some money," Fred says. "We have to think: How much more water can we reserve for ourselves?"

The Joint Powers Water Board is in a unique position because they're not competing against anyone else locally, but part of the reason they're building a reservoir is to be more secure. They're worried about what might happen if downstream water users in places with bigger populations start to claim their full share because the structural deficit might screw them over. It's much harder to get water back than it is to develop it initially. You can't cut off water to a major city once it's using it. That conflict point hasn't been tested yet, but with coming climate change and the looming lack of water, it could be soon. Fred says they're looking at using all their rights now so they can lock them in for the future, even if they don't know what the future might look like. Wyoming's population overall is shrinking, and it's unlikely that even boom times will dramatically change that in the long term, but it's fatalistic to think about not protecting the water for future use.

That's why Rock Springs and Green River are becoming increasingly vigilant. They don't ever want to be at risk, and they know the stakes are getting higher. The future is unsure, but at some point there will be a shortage. Transbasin diversions, drought in downstream cities, and a squeeze on the structural deficit all mean that they might not have water at some point if they don't develop it now. "Wyoming doesn't use all the water we have, because Wyoming doesn't have a water crisis yet," Fred tells me. "It's everyone downstream."

FLOWING UPHILL TO MONEY

We wake up in the still-damp grass on the riverbanks. I'm the early riser, so I'm in charge of mornings: boiling water for coffee, reinflating the boats when they've gone slack in the chill overnight, nagging Will and Dutch to get moving, and trying to keep track of river miles and how far we have to go each day. Many of the bends look the same and the river here is never rowdy enough to challenge us, so sometimes the hours glaze together. They're in charge of dinner and storytelling, and they stay up later than I do, tending twiggy fires and picking out constellations they've never seen before in the open desert sky. Even as we float toward the city of Green River, the range around us feels wild, with no signs of people except for the occasional power line and a few bridges.

In addition to downstream urban use, climate change is shrinking the water supply and changing how and when the water flows. Higher temperatures create more evaporation, which directly affects water supply.[1] Every year since 2013 has been hotter than the last. And the long-standing drought the area has been experiencing since the beginning of the century has exacerbated the structural deficit in the water system. Places like Green River are fearful because neither supply nor demand feels stable, and, just as I do when I'm paddling, when you're negotiating urban water rights, you always have to be looking downstream.

Some western cities are getting really good at water conservation. Las Vegas—surrounded by desert and arguably one of the most naturally water-disadvantaged cities in the world—reduced its per capita water use by 36 percent between 2002 and 2017, while gaining half a million more residents, by enforcing landscaping rules, spurring public education campaigns, and paying homeowners two dollars a square foot to get rid of their lawns.[2] But that's not the case everywhere—for example, according to the USGS, Utah uses more water per capita than anywhere else in the country, due to cheap water rates subsidized by bonds.[3] Because so much happens on a state-by-state basis, it can be hard to enforce equitable water use and reductions, particularly because each state is legally entitled to its share of water and can technically do whatever it wants with it. Once a place has developed its water, that development is nearly impossible to undo. That's why cities like Green River are trying to make sure they're not the ones left dry.

Individual states can, theoretically, pipe their allocated water to wherever it's most needed or most valuable, even if doing so drastically changes the nature of the river. That often results in water being diverted out of less populous, but wetter, river basins and into places where there's a greater human need, and a greater ability to pay. Transbasin diversions were first built here in the 1860s. They're based on that same landscape-altering sense of western manifest destiny that spurred the Homestead Act, and they're the reason places like Las Vegas can exist in the midst of a desert. That sense of unbalanced supply and demand is particularly obvious here on the Green, where the demand is low, and the supply is present. Looming proposed transbasin diversions are one of the biggest potential future risks to the Green.

I understand that urge to find a way to live here because I'm part of it. I'm an easterner pulled west by the promise of big ranges, wild rivers, and a landscape-scale reflection of the part of me that felt untamed. I'm not sure how much of that pull is nostalgia, a need for space, or wanting to feel like I had a stake in something unclaimed, but I'd been fascinated by my internal conception of the West for as long as I can remember. As a kid I ripped through John McPhee and Wallace Stegner, trying to absorb the geologic scale of the region through books. I came to Colorado for the first time as a teenager and promised myself I'd move there as soon as I could.

When I drove my loaded car out from New England at twenty-one, headed for the mountains, my life didn't become as vast and untouched as I'd assumed it would be from the East Coast. I became enmeshed in an everyday that didn't exist in the deep wilderness. But every once in a while, even after my eyes adjusted to the scale of the Rockies, I'd still get stopped by late daylight on the edge of a peak and get that crucial feeling of smallness.

As we curve down the river, I watch that same sense of scale hit Will: horizon-wide sunsets, masses of swallows, pronghorn bounding away from the river when they hear us coming, and, sometimes, just the emptiness. On our first day on the river we have a hard time keeping track of how far we've gone, and we keep misidentifying landmarks because things feel so empty.

I think there's a trap in loving that empty western possibility, especially when you don't live in it all the time. That's part of the deepening divide between urban and rural areas, and it's tied to how water is used. By being there, you're changing it, and by being removed from and romanticizing it, you're probably not seeing it change. I'm guilty of that. I live in a city now, and I don't get to see these landscapes on a daily basis. I have an idea in my head of how I expect them to look, but it's abstract and blurred, and now, being on the river, watching it every day, and talking to people who depend on it directly, I start to realize how out of touch I've become, even though I've been trying to pay attention.

Just downstream of here, water is diverted to the urban, growing Wasatch Front, the biggest population center in Utah, home to Salt Lake City and its sprawling suburbs. Most of the state's allotted Colorado River Compact water comes out of the Green and its tributaries through a complex series of transmountain diversions and reservoirs called the Central Utah Project.[4]

The CUP, which was designed to corral Utah's portion of the Upper Basin's water, is the largest water development project in the state and one of the biggest in the country. Water gets pumped out of the western tributaries of the Green—the Duchesne and White Rivers, Ashley Creek, and more—into a series of reservoirs and tunnels that eventually dump water out onto the west side of the Wasatch Range. Congress authorized the project in 1956, and completion of the Bonneville Unit, the final piece of the project, is scheduled for 2021. It's huge. The Bonneville Unit alone is a mesh of ten reservoirs and hundreds of miles of tunnels, canals, and drains. Its development has never been straightforward. In 1978, President Carter almost cancelled it in an attempt to cut back unnecessary water projects, and there's still an unresolved settlement with the Ute tribe about transfer of water rights from the never-built Ute Indian unit of the project.

The water that comes through the CUP's pipes has made Salt Lake City livable, and it's what makes water cheap there. "We bring 102,000 acre-feet across the divide from the Colorado River Basin to the Wasatch Front," says Gene Shawcroft, the general manager of the Central Utah Water Conservancy District, which manages the CUP. He says that the district uses only about 40,000 acre-feet currently, but that it's planning to grow into its full allotment over the next ten years, mainly through urban growth. He says 90 percent of that water is already contracted out for future development, and that the CUP is banking the rest for when it's needed. It's keeping the reservoirs as close to full as possible. That's the coming risk that worries places like Rock Springs and ranchers like Pat O'Toole. That growth is big and powerful.

Utah's current population is 3 million, and it's expected to double to 6 million by 2060.[5] Salt Lake City and its surrounding municipal areas, which are already some of the fastest growing in the country, are slated to grow significantly in the next few decades. Gene says that in planning for the future, the CUP is trying to figure out how to get water to those people.

In water management, city water encompasses municipal and industrial uses. In Utah, which has an exceptionally high per capita urban water use rate, city water accounts for about 15 percent of the total water used in the state, and that percentage is rising as populations increase, climate gets warmer and weirder, and farmland is converted to municipal uses. The state has put water conservation goals into place in an attempt to plan for the impact of urbanization on its water use, but the results have been marginal. Utah is still using water at almost twice the national per capita average rate, in part because there's no financial or social incentive to save it.

I think that's because of the same myth that drove me to the mountains, the baked-in historical assumption that the West is still empty, unlimited, and up for grabs. When the initial white settlers started overruling nature's seasonal hydrograph and making deserts livable, they changed that reality, but there's still a cultural context that says it's true. Cities are getting more efficient, but they're also getting bigger. You can't tell people where to live, and lots of people want to live in the Denver foothills and the deserts of Phoenix. That growth doesn't show any sign of stopping, and those people all need water. In the balance of water use, the agricultural slice of the pie is getting smaller and the urban sliver is expanding. The size of the pie remains the same, so the slices have to shift. Often, that means finding ways to transfer water from farms to cities, potentially leaving those farms dry forever and deepening the divide between rural and urban economies.

The Wasatch Front isn't the only place where transbasin diversions are a factor. There are thirty in Colorado, because most of the population of the state is on the eastern side of the Continental Divide. Water gets sucked out of the rivers on the western slope, which drain into the Colorado and Green, and ends up flowing east instead, limiting the amount of water that runs down to users in the Lower Basin. There are several proposed pipelines that keep getting put on the table when water supply seems low, so there are potentially more straws coming.

The biggest out-of-basin threat to the Green has come from a Fort Collins, Colorado–based developer named Aaron Million, who has tried for years to build a pipeline that would siphon 81 billion gallons of water a year out of the Green River Basin and send it over the Continental Divide to the Denver area, at a cost of $9 billion. The Federal Energy Regulatory Commission last shot his plan down in 2012,[6] but he raised it again in 2018 as drought crept in. On the western slope of Colorado, folks call it the zombie pipeline because it keeps coming back from the dead. Another similar proposal, from a group of city water districts in both Colorado and Wyoming, which called itself the

Colorado-Wyoming Coalition, was put forth in 2009. All such projects have been squelched so far, but people keep proposing them when it feels dry on the Front Range, because the Green still contains the biggest surplus of unused water in the Upper Basin. In the recent history of the Green, it's been sharky independent developers who've been proposing these diversions, not state or local water managers, but they're not idle threats, because at some point cities might need enough water that the economics could make sense.

Steve Wolff, the Wyoming state engineer's representative for interstate streams, says Colorado would have to sanction any diversion of its water rights, and so far, it's refused to do so at some step in the process each time Million or someone like him has proposed a pipeline, but there's always a push for more water development. "There are no more main stem reservoirs planned on the Green, but we have unused water, and there's always discussion about moving it," Steve says. "The sponsors of the project think there's traction. I can imagine if Denver keeps growing we'll see a little of that."

The threat is real enough—especially with continuing growth in the cities—that environmental groups go on high alert any time large-scale diversion projects are proposed. River advocacy nonprofit American Rivers listed the Green as the second most endangered river in the country in 2011, the last time Million put forth a proposal, because of those pipeline threats. A diversion would slaughter the water supply, harm fish habitat and recreation, and set the Upper Basin up for future shortages. Even though a diversion to Denver would be politically and financially difficult, it wouldn't be incredibly hard to build—it would simply follow I-80 to the east. Much more logistically complicated water projects have been built in the past. And it's legal, under the Upper Colorado River Basin Compact, to divert water from one state to another. It just takes Army Corps of Engineers permits, a Bureau of Reclamation contract, and approval from the governors. Eventually, it might come down to dollars. "We don't know what the cost is going to be in the future, but we do know some people pay a lot more for water today than it would cost to build the pipelines. Based on current economics and the price of water it works. When you look at the value of water and what water facilitates it makes a lot of sense to have water available to grow the economic engine. Utah's policy is we will grow." Gene Shawcroft, from the CUP, says about the future.

The Green River's status as some of the last unallocated water in the West accounts for the sense of vulnerability hanging over Rock Springs, and it builds in a layer of competitiveness. Wyoming will cling to its water storage, but if Utah or Colorado develops downstream water first, and then climate change creates a shortage, Wyoming might not ever get its full allotment.

But, like a lot of resource issues, this one can feel abstract and removed from the reality of the river until the water is gone and it's too late. It's hard to picture things outside of your frame of reference, and the scale of water use is hard to grasp, but it touches everything. For most of our float down into the city of Green River, Will, Dutch, and I get what feels like emptiness instead of urbanity. The Red Desert stretches to our east, covering almost ten thousand acres of grasslands and canyons. It's fraught in its current state of untouched starkness; conservationists are trying to protect it as habitat while energy companies are trying to lease it for drilling. We can start to feel the city coming as we get closer. For most of the trip we've seen only a few scattered fishermen, but as we approach the city we can see a few ranches, then the outer urban detritus: broken-down cars in ditches, then houses, then the slick green roof of the water treatment plant.

WHOSE RIGHTS?

In trying to parse the big picture of who gets what water and how that changes, I've been having a hard time wrapping my head around water rights. They're the root of everything, and they've been mentioned in every conversation I've had so far, but I'm still confused about how they're allocated and accounted for, and what's left to squeeze out. As I hear more about transbasin diversions and how urban centers might buy up agricultural water rights, I'm still unclear about how the system works, what the numbers are, and how states can legally divert water to a completely different river basin. Understanding these things feels important in understanding how much might be left and how precarious the supply might be. Water rights, for the most part, are governed by states, so before I get on the river with Will and Dutch, I track down the nearest state water office, hoping for answers. In Vernal, Utah, downstream from the Flaming Gorge Reservoir, I go to the Utah Department of Water Resources and spring my questions on Andrew Dutson, the assistant regional engineer.

I walk it, sweating from the summer that seems to have sprung up all of a sudden. I'm not even sure if I know what questions to ask, or if I'm using the right words, but I want to know how one might acquire a water right and what it takes to change them, so that I can understand how big remote cities can pull water away from ranches, and why places like these might be worried. Luckily, Andrew is slow talking and detail focused, exactly the kind of person you'd want to have managing the interlocking minutiae of how water rights are designated, accounted for, and guaranteed. He's slightly balding, in an orange T-shirt. Crayoned pictures from his kids plaster his office walls. I ask him what would happen if I applied for a water right, and he starts walking me through the process.

The state or the federal government can control water rights. Those rights can be considered private property, as they are in Colorado, or public property that users are permitted to lease, which is how things work in Wyoming. In Utah, for the most part, they're administered by the state, and they're tied to the land and the specific beneficial use. Most of the state's water is already allocated, at least on paper (what's known as paper water rights), although not all the allocated water is already developed and in use (commonly called wet water rights). According to projections from the Utah Division of Water Resources, in 2020, Utah will have about 200,000 acre-feet of undeveloped Colorado River water available for future use. By 2050, it will have 42,000 acre-feet. The figures are changing quickly—in 2000, it had 416,000 acre-feet.[1]

Andrew pulls up an online map that divides the state into regions by drainages and starts to talk me through the different ways in which those regions allocate water, through seniority, different kinds of beneficial use, and a series of proofs that the user must adhere to. After five minutes I'm already looping back on questions, confused because there doesn't seem to be much consistency, or a standard way of doing things. Every instance he mentions, from agricultural water rights to industrial ones, seems different and specific to its case. "We have the same general rules, but policies are different depending on which region you're in," he says. "The gist of it is that it's complicated, and ever-changing depending on where you are and what the availability is like."

The way we use water is fundamentally changing, too, and laws and management practices, because they're tied to decades-old regulations, aren't necessarily keeping up. As the states in the basin transition from predominantly agricultural economies toward growing urban, municipal, and industrial uses, water rights are changing, but there isn't any more water. The unused portion is shrinking, so uses have to change. Andrew says that cities and industrial users often buy

parts of other rights, or get temporary or fixed-time rights, which theoretically could be cut off. The process of buying up agricultural rights,
then diverting them for use elsewhere, usually for municipal or industrial uses, is commonly referred to as "buy and dry," and it's one of the
biggest and most common changes in water use. It can completely alter
an ecosystem, a local economy, and downstream flows, but it's legal
within the bounds of the water rights system, which is the overarching
framework. What makes sense legally doesn't always align with what
make sense ecologically, and economic benefits don't always follow
the water.

Andrew tells me that Utah started keeping track of water when the
state engineer's office was established in 1897. The Utah code, which
formalized the Doctrine of Prior Appropriations, was put into place in
1903, and when the Colorado River Compact was formed in 1922, Utah
started using the compact numbers as a baseline to gauge how much
water it had, and how much it had left to use. Utah gets 23 percent of
the Upper Basin's flow, or 1,725,200 acre-feet, based on that 7.5 million-
acre-foot inflow that was estimated in the compact, but a lot of times
the inflow is less than that, so the numbers the state is using to start
with aren't fixed. Which means it's subtracting from an unknown number. First, there has to be water available. "Generally, people come up
with where they want to get it from and what they want to use it for,"
Andrew says. Then, the intended use indicates how much they can use.
There are specific amounts associated with irrigation use, for instance.
But there's no standard industrial amount, or municipal amount, and
there are some cases in which historical precedent sets the amount.
The Jim Bridger Power Plant, near Rock Springs, for example, gets
120 cfs, which is three times the amount the JPWB gets to supply the
whole county.

Once a right is approved, the user has to put it to beneficial use, and
then submit documentation to the Department of Water Resources to
prove that they're using it in the way they said they would. Confirming that can take years, which is why there's still a lot of slack within
the system. For instance, Andrew is currently working on confirming
the Bureau of Reclamation's federally reserved right to water from the

Flaming Gorge Reservoir, which was filed in 1958. "I'm not done yet because we want it to be final when it's done," he says. "It's a lot harder to fix it after the fact."

All water uses in the state go through this channel. It's possible to trace rights from before the compact, and even from before Utah was a state, to the earliest mining and ranching rights. Understanding Utah's process gives me a baseline, and a little bit of a sense of how rights are currently doled out, but that process doesn't necessarily apply across the board. Exports of water out of the Upper Basin in Utah, for instance, are predicted to almost double in the first half of the twenty-first century. That will have a big bearing on the amount of water that will be available, but it's within the state's rights to export that water.

Every state along the Green has a state water plan, which spells out how it manages its part of the river. Colorado released its plan in 2015,[2] after spending 30 months listening to stakeholders in the state's nine river basins. Utah made a statewide plan in 2000 and is working on a new one, one river basin at a time. Wyoming's plan was revamped in 2015, and Nephi Cole, the governor's water policy advisor there, says that planners took into account all the ways they saw water use changing in the state and tried to predict how it might change in the future.

The framework of the compact encourages states to allocate all the water within their rights, instead of leaving any in the river for the ecosystem. There's very little legal incentive not to use up all the water, and by trying to lock up all their water for potential future use, Wyoming's plan hews to that law. Legally, they might as well allocate every drop. In some states, baseline flows for fish aren't considered a beneficial water use, so a waterway can be sucked completely dry. Water replenishes itself to a degree, but once it's pulled out, stopped, or diverted, that changes the course of the flow. Cities across the West are facing a moral dilemma: Should they use up as much as they can to ensure that they have water for the future? Will they be bullied or priced out by bigger cities? If they try to conserve, will they lose out in the long run?

As I sit here with Andrew, it seems as if water rights are allocated and used in a vacuum, far removed from the way the river actually looks and functions. They're essentially currency. To abide by the laws, you have to block out how much you might ever want to use, then use it all up. And from Andrew's end, if the state is trying to spread out water as equitably as possible and give as many people as possible a fair crack at it, it makes sense that they would allocate all of the water that the state has in its rights. You can't live without water, and here, in the thirsty high desert, where it's already hard enough to eke out a living, access to water feels like another source of disparity.

When I leave Andrew's office, I have a better sense of what a water right is and where it comes from, but less of a grasp on how the system as a whole makes sense. It's both rigid and flexible, and that creates friction, because none of it exists in isolation. There are a lot of different, broadly polar ways of looking at a river: you can think of it as a connected habitat that needs constant attention, or as a plumbing system that should be used until it's maxed out. The water rights system plays to the second view. It made sense when it was created, and a standardized way of apportioning water with hard parameters was necessary. It plays into the idea of western individualism, but in a connected water system, actions have consequences all the way down the river.

The next week, on the night Will, Dutch, and I float through Green River, the rodeo is in town, and we decide we need to go. We wash off a week of river scum, put on our best approximation of rodeo gear—a snap-button shirt for me; cargo shorts and an American flag–printed T-shirt for Dutch—and head to the rodeo grounds. We get back-row seats in the bleachers, next to a young mom in cowboy boots and shorts bouncing her tiny baby. Will climbs the stairs with three Budweisers in his hands, the constellation of stadium lights clicks on, and team roping starts. The crowd, which skews young and family oriented, is peppered with casts, knee braces, and limps. A kid with tiny, stiff, straight-leg Wranglers and a gigantic hat wins the mini bronc competition. He walks back to his parents in the stands, holding his new belt buckle out in front of him lock-armed, like it's a breakable treasure. I have spent the past few weeks paddling, using my body every day, but this, somehow, feels more physical and visceral.

What I think I know and how I feel are in conflict. I've been reading about how rural America is hollowing out and dying, but tonight, under the buzz of the fairground lights, I am jealous of the people who live here. I know the rodeo is a show, but it feels real and tough and beautiful, tied up in history and community. I want to be a part of it, but I can't. We're not from here, we put on costumes, and we'll be gone in the morning. Maybe the rodeo is an outdated myth of the American West, romanticized and simplified into pageantry with all the gritty parts cut out, but as we watch the roping, then the riding, an ache sets in for me. When the cowgirls run their horses down the cloverleaf's straightaway, standing in their stirrups, curling-iron waves blown back behind them as they gallop home, I want to be a barrel racing queen. It feels like all those stories I've told myself about beauty and ruggedness and western possibility are coalescing. I'm sad for the pasts I never got to have. "I had no idea places like this actually existed," Will says again, as the sun slopes close to the horizon.

This specific way of life is getting smaller, and harder to see a slice of, as people move away from the rural West. That's true of water, too, because the question of how water is used is really the question of how people live and make their livelihood. From a strictly economic view, it makes sense to send it to dense, efficient cities where users can pay more. But as towns like this empty out, and the agricultural and extractive industries of the past shrink, there are people still living here. They need water too. And it's hard to think of them as real until you know they exist and know their stories.

The rodeo is pageantry, but it's also sport. During the saddle bronc competition, a cowboy falls, twisting, under the thrashing horse. He doesn't get up. Then he still doesn't get up. The rodeo clowns make a wall around him with their bodies, and the ambulance circles in under the lights. I have a lead-gut feeling that we've just watched someone die. The announcer patters emptily, trying to drown out the silence, telling stories about how good a rider the cowboy's girlfriend is, trying to lighten the mood. Eventually the cowboy stands, stumbles a little, waves. The crowd, which had gone silent and grip lipped, claps as he's loaded into the ambulance, and the night moves on. The rodeo princesses resume making the rounds of the stands, signing pictures of themselves. One of them takes a high-speed lap around the ring on her horse, flashing silver and turquoise, and waving an ExxonMobil sponsor flag, a reminder that the myth of the cowboy is based in truth, but gilded over. It's a story about independence, toughness, and living close to the land that we tell ourselves from afar when we don't see people's real lives clearly enough, or patiently enough, to understand.

The next day, back on the river, we float out of town toward Flaming Gorge Reservoir. The river gets flatter and browner as we move into the top of the lake, and we have to paddle harder to push ourselves downstream. I can't shake the image of the motionless cowboy, and the way the rest of the crew silently formed up around him, making a wall so we couldn't see anything bad happening.

DAMS

6,940 CFS

CLAIMING AND RECLAMATION

I drive back north through the swooping brown hills above Green River, and the edge of the Fontenelle Reservoir comes into view like the mirage of an oil slick on a hot highway. The color seems wrong. Should there be pelicans? Who would put a body of water in a desiccated place like this? The river drops and slows, puddling back on itself. The soil is so dry it cracks when I walk down toward the water.

When we'd paddled from the Fontenelle Dam into Flaming Gorge Reservoir, crossing the state line into Utah, we'd seen the river change from rushing cold and clear below the dam to muddy and slow in the backwater of the second reservoir. The transition was obvious as we moved downstream, but I wanted to know more about how dams changed the dynamic of the river, so I backtracked to the Fontenelle Dam—the first one on the river—to see if I could talk to someone who knew.

That morning, I'd cold-called Kirk Jensen, who runs the Fontenelle Dam and Power Plant, to ask if I could see the inside of the dam. "Is now good?" he wanted to know. "We'll open the gates for you." The chain-link fence was unlocked when I drove across the empty, single-lane road on top of the dam. I wound down a dirt road to the blocky operating house at the bottom of the downstream side.

Fontenelle is a utilitarian dam. It has square corners and lacks any kind of architectural grace. When Kirk answers the doorbell and walks me inside, I get the feeling that a random visitor is a rare occurrence. He leads me into the narrow, humming control room, which is lined

on both sides with analog dials and switches. It feels like we're inside a submarine, monitoring water pressure below the surface.

I follow him down a set of skinny stairs to look at the guts of the power plant. Fontenelle has a single Francis-style turbine, a kind he says is often used on rivers with large flows and dams with steep drops. The turbine creates a small amount of power—10,000 kilowatts—but Fontenelle mainly operates to store and supply water and to keep its reservoir, and Flaming Gorge Reservoir downstream, at certain levels. The amount of water pushed through the dam is based on river flow and demand downstream.

According to the World Commission on Dams, "the United States is home to approximately 6,575 dams at least 15 meters tall, second globally to China's 22,000, and well more than twice as many as the rest of North America, Central America and South America combined."[1] The ten largest dams in the United States—and two-thirds of U.S. hydroelectric capacity—are in the West, and they were all commissioned by the Bureau of Reclamation.

The dam building boom began in 1902, when Congress passed the Reclamation Act, establishing the Bureau of Reclamation, called the Reclamation Service at the time, within the Department of the Interior. The act required water users to "repay construction costs from which they received benefits," creating a large federal funding stream for water projects.[2]

The Bureau of Reclamation was initially part of the USGS, which had done much of the mapping and exploring west of the Mississippi in hopes of figuring out what parts of the frontier were habitable. Early Reclamation employees were given the directive to "reclaim arid lands for human use"—in other words, to build water storage projects for irrigation and then later for cities. The idea, based in the science of the time, was that harnessing water would make the desert livable—that people could reclaim it from wildness. In tandem with the Army Corps of Engineers, the bureau built 180 dams across the West. By the late 1980s—and after $11 billion of investment—most of the dam building projects had petered out, but the bureau had fundamentally changed a significant portion of the country's rivers.

The bureau's overarching goal, like that of the Homestead Act, was settlement, taming the wild western landscape and making it profitable. But it set up a double bind. Dam building became a moneymaking venture as well as an infrastructure build-out. There were benefits. The bureau provided jobs in the wake of the Great Depression, and it did what it set out to do: it provided a large-scale plumbing system for expanding agriculture and growing cities. But that growth came at the wide-ranging expense of the surrounding native ecosystem, and many of the dams weren't necessary. They were built in areas where farming wasn't tenable, or where population centers were far away, because local politicians wanted federal funds. The dams disrupted the seasonal, snow-spiked hydrology of the rivers they plugged and changed everything downstream. According to the theory posed in defense of dam building, dams limit the risk of water shortages because they build in stability and storage, but they also add vulnerability to fragile natural networks.

Dams have a complicated history and a fraught future. The bureau says its projects provide agricultural, household, and industrial water to about one-third of the population of the American West, and that's true. We've staked life in desert cities on their ability to deliver water steadily. Water storage is a big part of the reason people can live in areas where it doesn't rain, but that benefit comes with big costs. Large-scale dams have also fractured ecosystems, stopped sediment transfer, changed the temperature and chemistry of river water, and blocked fish passage. Evaporation from reservoirs is a significant, consumptive water use that can eat up to 10 percent of the river's flow in dry years. Even in average years, Lake Mead, one of two major storage sites for the Lower Basin, is shrinking to levels that limit its ability to provide both water and power, which could drastically change life in the Lower Basin states.

The Seedskadee Project, the official name of the Fontenelle Dam construction project, was authorized under the Bureau of Reclamation's Colorado River Storage Project Act of 1956. The act initiated construction of a number of large dams, including Glen Canyon and Flaming Gorge as well as Fontenelle, with the intent to manage and mitigate risk in the upper Colorado River and to control flows between the Upper and Lower Basins. By the time Fontenelle was built in 1963, it had already become clear that there was less water in the river than was allocated in the Colorado River Compact.

There's a constant thrumming fear around the lack of storage in the Upper Basin, and because of that anxiety, Fontenelle is at an interesting crossroads. It was initially built to supply water for irrigation, but that never happened. The area around it is too dry, gravelly, and cold to grow crops. A test farm run by the bureau after dam construction started quickly proved that, so these days, Fontenelle is a storage reservoir, which helps guarantee that Wyoming can get its designated 14 percent of the Upper Basin's water. But because of Fontenelle's original purpose, which would never have drawn the reservoir very far down, the armoring that was installed to prevent the earthen dam from eroding when surface waves hit it—called riprap—extends only halfway down the dam. Because of the resulting threat to structural integrity, Kirk can never let the reservoir get too low, so the water in the bottom third is inaccessible. In February 2016, a bill passed the U.S. House, and was introduced in the Senate, to expand storage in Fontenelle by bringing the riprap down to the bottom, which would give the state access to all the water in the reservoir.[3]

The Fontenelle storage expansion is part of Governor Matt Mead's Wyoming Water Plan, which calls for ten storage projects to be built by 2026,[4] guaranteeing more water storage for Wyoming and securing more of their compact water. The state wouldn't build any big dams on the main stem of the Green; the small-scale storage projects, most of which are slated to be off-channel, would just change when and how much water would be coming downstream. Wyoming is trying to reframe history to plan for potential future needs.

When Kirk and I walk out onto the deck above the turbine, a tempest of whitewater flows into the river below us. With the laconic practicality of a government engineer, he explains what the course of his year looks like. Spring, which comes with a rush of snowmelt that needs regulating, is when things are busiest, because he's trying to balance runoff with downstream water needs and with filling the reservoir for the upcoming summer and fall. He says the reservoir is usually at its highest around July 4, and he thinks they'll fill it up this year. After an unusually dry winter, the spring has been wet and cold, and it's been raining more than normal. He says they're pushing 1,550 cubic feet per second out of the turbine right now. If the reservoir gets too high and they need to dump water, they can let out as much as 6,500 cfs through bypasses that flow around the turbine. If they release more than that, they start to flood out the fields around Green River. "We have to control it slowly," he says.

Humans have been damming rivers since the beginning of civilization, especially in arid areas. There are ruins of eight-thousand-year-old irrigation dams in Mesopotamia, and there are pre-Christian aqueducts in the Roman world that are still in use today.[5] When modern-day dam managers talk about reservoirs, they mainly talk about storage as a means for controlling the unstable balance between water supply and demand. It's the antithesis of a natural flow regime, but in a lot of places it's necessary. Water rights are irrelevant if you don't get your water when you need it.

In Green River, Fred Ostler told me that the city is highly dependent on Fontenelle, both to prevent it from flooding with spring snowmelt and to keep its supply of water steady though the year. Because the water treatment plant currently can't store much more than a few days' supply, it relies on the dam to release the water it processes so residents have something to drink. The river makes the city livable, but the dam makes it possible to live there year-round. "The general public doesn't know what the dams do. They'd be at the mercy of the rivers without them. These dams and this hydropower, to me, we need them out West," Kirk says, as we watch water pound out into the eddy below the dam.

After Kirk's tour, I head south, back toward Flaming Gorge, the next major reservoir on the river, tracing the section I'd already floated with Will and Dutch, glimpsing silver flashes of water in the dun of the desert as I drive. When I cross the highway near Rock Springs, into territory I haven't seen before, the landscape changes significantly. The geology tips up, exposing older rock. I drive backward through time,

into deep red, ancient Uinta Mountain Group rock. After a month of subdued desert colors, everything looks vibrant and mountainous, with pines instead of scrub and the deep green of the reservoir to my right.

Will and Dutch have one more night before they head back to DC. I catch up with them in Red Canyon, on the southeast side of Flaming Gorge Reservoir, and we hike out onto the ledgy cliffs above the water. Flaming Gorge is Moenkopi, Chinle, and Navajo Sandstone from bottom to top, striated and summer colored. The ninety-mile-long lake is all skinny fingers here, nearly black in the deep side canyon thousands of feet below us. When we stop on a ridge to peer down into the lake, there's a golden eagle circling under us, catching the thermals rising from the canyon walls, flashing in the light. The boys are summer-vacation giddy, jumping between the spindles of rock, willing themselves closer to the edge like kids. We traverse the rim until the light gets low, then find a spot to set up our tents.

On the edge, it's hard to equate the vermillion canyon with the utilitarian supply and demand I know Flaming Gorge and Fontenelle create. I understand it's an artificial landscape, but as the evening shimmers in I still get the kind of electricity under my skin that I associate with wilderness, the feeling that I'm in a living landscape, rangy and untamed. I don't know what it was like before the dam, or whether the contrast of the dark lake below makes the canyon feel vaster than it otherwise might, but it's the kind of beauty that changes my feelings, that makes it hard to be neutral and level-headed about water. Coming into the trip, I knew enough about reservoirs, dams, and storage to think about them as a necessary evil: man-made breaks in the ecosystem to be worked around, reckoned with, and taken out if possible. I was removed enough from the water system that I hadn't thought much about the value of a steady water supply, and I really hadn't thought I'd find this part of it beautiful.

Almost every part of the landscape I've seen so far had been altered, so the value judgment seems like it has to be on the kind of change, not whether it's changed at all. How do you do the most good, what is the most fair? Kirk's words stay in my head as we feed sticks into the campfire, willing our last night together to last longer. "People don't realize how much we need this," he'd said.

In the morning, the guys track back toward the Salt Lake City airport, and I head to Flaming Gorge Dam. The road loops around the lake, and at the skinny southern tip its two lanes trace the rim of the dam, which seems like an impossibly thin fingernail of concrete. On the upstream side, there's a marina and a museum, and I stop there to take a tour. Inside, in the air conditioning, I wait next to a timeline of the Bureau of Reclamation's history before I'm escorted through a metal detector by a khaki-shirted security guard. Once they know I'm bomb free, I'm allowed to walk out along the upper deck of the 502-foot-tall thin arch dam. After a succession of dirtbag days on the river, I'm suddenly in starchy, rule-filled summer vacation land. A man with three little kids keeps quizzing his sons on what they're learning. An older woman is taking photos on her iPad. Gayle, our tour guide, says visitors used to be able to take the tour alone, before 9/11, but now we're stuck with her. She's grandma-aged, with shellacked nails and glossy lips, and she launches into the backstory of the dam. It was built starting in 1958, in the first phase of the Colorado River Storage Project. Concrete was poured for two years. President Kennedy turned the turbines on three months before he was assassinated. The ninety-one-mile-long reservoir, which holds 3,788,900 acre-feet of water, took twelve years to fill. Gayle tells us the water in the reservoir is currently at an elevation of 6,032 feet, and that 6,040 is considered full. I'm not usually floored by large engineering projects—Fontenelle hadn't moved me much—but something about the scale here makes me feel uncharacteristically patriotic and proud.

The dam feels stouter when I can look over the upstream edge to see the press of the lake behind, and as we climb up and glimpse the downstream side, it's eerily beautiful. Wires string up from the power station below, and there are buzzards sitting on the crossbars of the transformers. Hundreds of feet down, the bypass tubes are firing, shooting out solid streams of white, frothing the emerald water below. Gayle lets us inside, and we take a cargo elevator down to an unnumbered floor where we can see the tops of the three turbines. It's a stark contrast to the inside of Fontenelle, where the surfaces had a layer of tools and empty Mountain Dew bottles. I can't see any mechanisms aside from the analog monitoring system, and everything is extra clean and tourist friendly. The turbine tops are glossy blue; the floor is a footstep-free gray.

We walk out onto an observation deck right above the river and watch the surge of water shooting through each of the two bypasses, sending compression waves back up the channel toward the toe of the dam. There are monster rainbow trout circling the platform, pocking the surface when they jump. The tail water of the dam is a gold-medal trout fishery. Sediment settles behind the dam, and the intake for the turbines pulls in clean, cold water from the middle of the water column, which the fish love. Below the deck, the trout are smart and acclimated to people like us. They know to look for food. For a quarter you can buy pellets to throw down to them, and they circle, smashing fins and thrashing for the food. The tour of the dam finishes there. Gayle leaves us on the deck with our glimpse of the power of the water it holds back.

The Colorado River Storage Project, which consists of sixteen dams in the Upper Basin, is one of the most complex water projects in the world.[6] After the tour, I slide through a back door in the visitor's center to meet Kirk's boss, Steve Hulet, the dam manager of the Bureau of Reclamation's Flaming Gorge field division. His office is filled with shiny wood and his desk looks out onto the dam. Steve, who has a long history with the bureau and a slow, southern way of talking, is in charge of managing how much water the dam releases. It's a complicated game of balancing inflows, runoff, storage, downstream needs, power generation, and habitat management.

Forecasts from the National Oceanic and Atmospheric Association (NOAA), which Steve relies on to decide how much water to release, had been predicting slightly below-average water supply for the year, but this spring, late-season rain spiked the inflow, sending more water than expected into the reservoir and changing the release plan. To avoid overtopping the dam, he's currently letting out 8,600 cfs. That's the maximum they can release without sending water through the spillway, which they try to avoid using at all costs. The turbines at the dam can handle only 4,600 cfs, so they're bypassing, and letting water out around the turbines as well, which Steve says he hates to do, because it feels like he's losing revenue from power.

Although Flaming Gorge generates 500 million kilowatt-hours of power each year—enough to power fifty thousand homes—power generation is secondary to water-level management here, as it is at Fontenelle. Across the West, where most of the big rivers are dammed, 20 percent of the electricity generated comes from hydropower, but the numbers are skewed by states like Washington, which gets 75 percent of its electricity from hydropower generated at places like the Grand Coulee Dam on the Columbia River, which sees flows that dwarf the Green's.

By midsummer, when the snowmelt has subsided, Steve says they'll be steadily releasing 800 cfs, the minimum flow federally mandated for endangered fish, then spiking water releases when the Western Area Power Administration, which markets the power the dam produces, asks them to, because more water equal more electricity. Sometimes, in really hot summers, they'll surge more water through the power plant's turbines twice in a day, once in the morning, when people turn their air conditioners on before they go to work, and then again in the evening, when they get home, to meet the demand for electricity. He says they try to avoid it because variable flows are hard on the fish and the commercial fly-fishermen who run trips right below the dam, but that it's also hard to keep everyone happy.

Initially, the bureau's work followed the path of John Wesley Powell, the second director of the USGS. Powell had a broad but largely self-taught knowledge of mapping, hydrology, and ethnography. He was iconoclastic and often right. He and his crew of nine men were the first white people to see all of the Green River from Green River, Wyoming, down. Once I hit the city of Green River I'd begun following their path, although it looks significantly different now that the dams are in place. Powell's trip was part of a highly ambitious survey that became the baseline for mapping the desert Southwest, one of the last big blanks in the map of the United States. The ten men put three wooden boats

on the water in a Green River park—the same one where I'd met up with Will and Dutch—on May 24, 1869. They floated down past the Green's confluence with the Colorado River, then through the Grand Canyon. They lost one of their boats, most of their provisions, and four of their men along the way. Powell went back to the canyons two years later for a second survey, and became the head of the USGS in 1881, but that first trip was pivotal in land use planning, both in his time and for decades to come, because of his careful analysis of river basins. We still use his framework. I think about Powell and his first crew a lot as I paddle downstream, especially once I get below the Flaming Gorge Dam into the deep red canyons of Browns Park and through the Gates of Lodore. I'm there at almost the same time of year he was. In some places our days line up, and I can imagine his crew testing their boats, the way I do mine, in bigger and bigger water as they moved downstream.

Powell named most of the geologic features along his trip, including Flaming Gorge, which he hit on May 27, three days after he and his crew put on the river. They hadn't seen much of the river yet, or paddled through any significant rapids, and the grandeur of places like the Grand Canyon was still to come. They came into the gorge in the evening and were struck by the late-day light reflecting off the red rocks. Powell labeled many of the geologic features he encountered more practically—he wasn't big on adjectives—so I'm guessing that when he floated into the Flaming Gorge he was hit by the same sense of awe I was.

I'm also guessing about what things looked like to them because most of what Powell saw in Flaming Gorge is now under water, submerged by the dam. Ashley Falls, one the most dreaded stretches of rapids on the river before the 1950s, is now deep enough below the reservoir that it doesn't even swirl the surface of the water. There were towns here, too, like Linwood, which was flooded out as the reservoir filled. We're constantly stacking histories on top of one another and telling new stories.[7]

Things change significantly below the dam. Tomorrow, I'll get back on the river and paddle through a section with class III rapids that's commonly called the Aquarium. Those rapids gave Powell's crew a hard time, but the river looks completely different now than it did to them. It's almost perfectly clear, with all the sediment trapped in the reservoir, and it's become a popular fly-fishing section. Like Powell, I'll have my first sustained test in those rapids, and they'll be the first big water I'll paddle alone.

It's almost impossible not to look at history through a present-day lens and blame the environmental corrosion of the river corridors on bad past decisions. But I try to give Powell and the Bureau of Reclamation engineers of the past the benefit of the doubt, and assume that they were trying to do good, trying to provide for the future in a time of deep uncertainty, just the way we are now. I'm struggling with that, with knowing what's natural and what's man-made, and what makes things better. I don't want to lose the spark of the canyon in the logistics of thinking of the river as a pipe to people.

To camp, I head to Antelope Flats, just upstream from the dam, where the reservoir twists and winds around the river's former curves. The canyons are spired and deserty, and the cliffs stretch up from the water, tall and buff-colored. There's a beach peppered with groups of campers. They've brought Seadoos and pontoon boats, camping tents and dirt bikes, and there's a bass-heavy pop-country thump coming from somewhere. I hadn't been planning to, but now that I'm at the reservoir's edge, I figure I should swim. I won't see still water for a while after this. I pull away from the other groups, strip down to my sports bra and shorts, and wade out into the murky chill. The water reverberates with engine ting and tastes like gas. I am peevish in the beauty. I have to remember that water means different things to different people. I lie back and float, toes trailing the sandy bottom.

AFTER THE DAM

In the morning I drive down to the Flaming Gorge boat ramp, beneath the concrete curve of the dam. They're releasing even more water today, and I can see the spray flipping up as we snake down the winding road. At the bottom, the water is thundering out; huge white compression waves pulse down the channel. It's menacing, made even more so by my aloneness. My aunt has come to drop me off and help me shuttle my car, but at the gate we learn that if you're not with a fly-fishing outfitter you can't park at the base of the dam, even for a few minutes, so she drops me on the side of the road with my pile of gear, gives me a nervous hug, and then drives back up the spray-slicked road, leaving me alone on the pavement with a deflated boat, in front of the biggest rapids I've yet to run. The seven-mile stretch below the dam is referred to as the A section, one of three in what's commonly called the Aquarium, which runs thirty-eight miles into the open valley of Browns Park. It holds some of the best fly-fishing in the country, and this morning, even though the water is up, which makes for tricky angling and paddling, it's crowded with commercial trips. Drift boats line the ramp, and customers mill in their khaki outfits and type V PFDs, taking pictures of the dam. I drag my boat to the edge of the water and start to rig it. This is the first time that being alone feels like drawing attention to myself. In the grip of anxiety about the coming rapids, I feel more isolated because there are people around.

Part of me wants to ask one of the guides if I can run through the rapids with them. I have a map, but I don't have a great sense of what's downriver, aside from Red Creek Rapid, which has been giving boaters trouble since Powell first paddled it. I get some curious looks, but no one asks what I'm doing, gearing up my tiny boat without a fishing rod or a companion. I chicken out about asking for company, and I'm self-conscious enough that I wait until most of the trips leave before I carry my raft into the water and kick off into the current, letting the leading edge of the boat catch the swirl of the eddy line.

This stretch feels different from anything I've paddled so far. The iron-red walls scrape the sky, and the water is clear green and shot through with light. The river is almost up to the cliff's edges, lapping at the red rock. A sign says there's a hiking trail on the left bank, between the tall pines and the bright new Russian olive trees, but the swollen river has submerged it. The canyon feels pristine to me, even though I know it's not. I know that the water should be cloudy and that there shouldn't be trout. If the water is high it should be because of runoff, not reservoir overflow, but the dam alters everything.

Steve Hulet, whom I met yesterday, runs the day-to-day dam operations, but the big-picture planning for Flaming Gorge releases comes from Heather Patno, a quick-talking, no-nonsense hydraulic engineer at the Bureau of Reclamation. She's the point person for complaints about water releases, which come frequently.

Water is inherently slippery. It's constantly in motion, which makes its management tricky. There are obligations downstream and pressure upstream, and a lot of the time the two don't align. Heather is the connection between them. She's in charge of prioritizing power and water needs, figuring out how much water to release and when to let it go, and making sure that as many people as possible are happy and safe.

Heather has some hard legal guidelines to abide by, including the Upper Colorado River Basin Compact, which mandates how much water has to end up downstream, and a 1992 Biological Opinion on the operation of the dam issued by the U.S. Fish and Wildlife Service,[1] which delineated federal water rights for endangered fish species, a major and often controversial water priority. The Western Area Power Administration (WAPA)—which is a part of the Department of Energy, and which serves fifteen states—sells the power Flaming Gorge creates, and it gives her directives for power generation based on air conditioning needs in Phoenix and other urban uses.

Like any large-scale energy source, hydropower has complex upsides and downsides. It's relatively renewable, and after a dam is built it's essentially emission free. Unlike other renewables, like solar or wind, it can be created on demand—you can turn it on whenever you want. And you can generate it locally, so you don't have to import oil. But to create hydropower, you need some kind of elevation change in the river, and for that, you need a dam. Dams fundamentally alter the food chain, flow, and function of a river. In many places dams do more harm than good, and it's nearly impossible to calculate all the associated costs and benefits before they're built, especially if you try to take into account loss of habitat and other long-term environmental impacts. That was one of the big flaws of the dam building boom of the twentieth century: when the Bureau of Reclamation built its dams, it didn't usually take species migration, or sedimentology, into account. We need power, but we also need sustainable rivers to guarantee that we have water in the future. Those endangered fish flows, for instance, often get targeted as anti-development, but in a lot of ways they're a safeguard that maintains a semblance of natural river function.

Utah is a huge producer of energy, for the state and for the rest of the country, but only about 2 percent of the in-state energy is generated through hydropower.[2] Of that, the majority comes from Flaming Gorge Dam. WAPA manages 40 percent of the hydropower generation in the western and central U.S., including the power created by Bureau of Reclamation and Army Corps of Engineers dams.

Because Flaming Gorge operates within the Colorado River Storage Project, it also has obligations to send water downstream into Lake Powell and Lake Mead to guarantee the Lower Basin's share of the compact water. Both reservoirs have been struggling to stay at tenable levels. The beginning of the twenty-first century was dry across most of the West, and both Powell and Mead have hovered at dangerously low levels, close to shutting down power generation and triggering downstream cutbacks in water use. Because Flaming Gorge is high up in the river system, comparatively small, and in a less dry climate, it hasn't been hurting as much as the downstream reservoirs, but part of Heather's responsibility is supplying Powell and Mead with enough water to keep them from dropping down too far. It's a constant guessing game with moving parts.

She also tries to make sure that flows below the dam stay within a range that doesn't flush out the downstream farmers who grow crops on the banks or the fishermen trying to find trout in low eddies. Starting in 1993, as dam operators were dialing in flows that would support fish populations in compliance with the Biological Opinion, they also formed a stakeholders' group, the Flaming Gorge Working Group, to give a voice to everyone affected by the changing flows. Heather sends them hydrology forecasts daily and warns them when the dam is going to increase or taper releases.

That information is particularly important for people who don't have a legal stake in deciding the flows but whose livelihood depends on the river, like farmers, rafting companies, and commercial fly-fishermen. The elasticity in the system messes with their businesses, so they want to know how much water is coming and when. Quickly ramping up river flows flushes trout out of eddies and floods farmer's fields downstream. Heather tries to keep them in mind, and to keep the conversations open, but there's often acrimony among these groups, who feel their voices matter less. And truthfully, in the sphere of decision making, they do. Those people don't have water rights, or any legal obligations. Sometimes reservoir management or power generation wins out, which creates anti-government sentiment among the less powerful factions.

As I get deeper into the canyon, the river feels swollen and taut in its banks, roiling with the kinetic energy of water flushing downstream. I am shaky and tense through the rapids, fighting to stay in the smooth tongues of fast current, overcorrecting to avoid every rock. I sing to myself through the wave trains to try to relax, even though I can barely breathe, and I fight the rail-grabbing swirl of the eddies harder than I need to. Rapids are all formed the same way—water, pressure, and rocks make waves, pourovers, and holes—but water levels change the character of a river. Here, it feels like everything is compressed and powerful, wound tight. The water churns even when it's not pushed into rapids. It's hard to tell how fast I'm going, especially on a river I've never seen before. Some of the fishermen I saw on the boat ramp are fishing in pockets of quiet water along the banks, but I flash past them in an instant, too fast to even wave.

I white-knuckle through roller-coaster waves and grabby holes, willing my little raft to stay upright. Aside from the rapids I ran with Mike and Ben, the river has been relatively flat until now. This is the first time I'm testing myself in any kind of sustained whitewater alone, and every wobble feels fraught. My heart speeds up every time I can't see past the flat of a horizon line, and I'm unsure of every drop. Small-scale, nontechnical class III rapids like these are referred to as read-and-run—the idea is that you can figure out where you need to be as you paddle through—but I stop and scout often, second-guessing myself as I search for the downstream vees, trying to find my line.

I pull into the Little Hole boat ramp at the end of the A section, still clenched, but proud, more confident in what the boat can do and what I can do in it. I stop, pee, eat a granola bar, and chat with three older men loading boats as I wait for the adrenaline to pump out of my chest. I'm still shaky. They're done for the day. Most fishermen just paddle the A section, because it's known for having twenty thousand trout in every mile, but fish density peters out as you get farther from the dam. They drive off; I slick myself with sun block, and head back out into the rangy, low-walled eight-mile B section.

It's hard to know what's coming. Heather's job is to predict the future based on the past, through the lens of a changing climate. Planning for the spring release cycle starts the January before, when NOAA begins publishing monthly forecasts for the unregulated flows coming down the Green and Yampa Rivers. Heather uses those forecasts along with a modeling program that indicates how much she should be letting out of the reservoir to keep it as full as possible without overflowing. She runs a simulation based on 107 years of historical data to estimate what the water will look like in the coming years.

But history isn't a direct predictor of the future, especially as global temperatures increase. "We've worked with scientists and the Intergovernmental Panel on Climate Change to account for variability under climate change," she says "It's the variability that is really significant. It's really important as reservoir operators and water managers to understand that we're looking at long-term trends. We're operating on whatever hydrology we get. Climate research is showing a shift from snow to rain. It's hard to manage for. We have to be prepared to manage water in wet scenarios as well as dry ones."

That variability makes it ever harder to predict what's going to enter the system and how precipitation is going to fall. This year, the wet spring brought 400 percent of average precipitation. Ashley Nielson, the NOAA hydrologist who's in charge of the forecast for the Green, says that that kind of weather pattern confuses both their long-range and short-term forecasts. "There are years that are trickier than others," she says. "The model doesn't do as well in extreme wet or dry years."

Ashley uses data from the USGS stream flow gauges and from the SNOTEL snowpack monitoring sites in the Wind River Range. She looks at where the precipitation is coming from, then models it against thirty years of data to anticipate what the water supply will be. She says they're starting to use climate data to help shape the forecasts, but they never would have forecast what they're seeing this year. "You don't forecast 400 percent of average," she says.

That precipitation spike means that they've opened the bypass at Flaming Gorge Dam and let the reservoir water come rushing through. It also means that I've flushed through the A section fast. What I assumed would take me most of a day has flown by in an hour and a half, and my hands are numb from clenching my paddle. I'd seen dozens of boats below the dam, but as soon as I float down below Little Hole, where the canyon starts to open up, I'm alone. The B section is mellow except for the biggest rapid in the Aquarium, Red Creek, which comes as a rude, churning surprise around a bend in the river. When I'd stopped at Trout Creek Flies, a fly shop in the dam-side town of Dutch John on the way down to the boat ramp, Steve Habovstak, the manager, who looked like he'd just drifted in from Baja in his flip-flops and deeply unbuttoned shirt, told me that a fishing guide had lost his boat in the rapid earlier that week. It caught an edge, swamped, and disappeared below the surface. A whole drift boat, gone. He said they figured they'd find it when the water went down, but there was nothing they could do until then. I can't shake that image of a boat being pinned and sucked under as I pull over on the river's left side to scout. No one would know I was missing for a while.

There aren't many slow-water eddies to catch because the river is raging, so I jam the boat into a sandbar above the tamarisk tangle, where warm, ochre-colored Red Creek runs into the river, and jump out quickly. The tamarisk, a knotty-rooted invasive, which is tough to get rid of and almost impossible to move through, makes it hard for me to scout. Fine-grained sediment from the creek forms a sandy island between the rapid and the shore. When I try to cross over to the sandbar to get a look at the rapid, I am immediately crotch deep in moving water, grabbing at tamarisk stalks to try to keep vertical. "So this is how it ends," I think at one point, trying to grip the bottom with my sandals. The banks are thick enough that I can't see anything without getting into places where I think I might get trapped. I don't know what to do, I have no one to ask for advice. I have to get downstream, and my heart rate is rising.

I clamber back to the top of the bank, sweating and scratched up, and decide to portage around the rapid, despite the nettles and brush that choke the banks. As I wallow back up, hot and frustrated, I second-guess myself, change my mind, and decide to try to run a sneak line in the narrow tamarisk-choked channel to the left of the island. It looked like it might be scrapy and skinnier than a boat in some places, but I'm exhausted and low on judgment, so I decide it will be less work than walking around, and less scary than running the unscoutable gut of the rapid. I push off and sneak through the boat-wide gap next to the island, scraping the bottom. It is just as narrow as I feared, and the boat drags and catches. My paddle tangles in the reeds, yanking my shoulder. I have to pull myself free from an arm of tamarisk at the end, but then I'm flushing out, pushed back into the main channel. By the time I catch my breath the river has bent again, and I can barely see the rapid from below.

I think about Powell again after I make it through the rapid, as I let the adrenaline shake out of my arms. When his party came through, they tied ropes to their wooden boats and lined them around the biggest rapids, avoiding obvious dangers and carefully gauging their level of risk as they pushed into the unfamiliar, trying to use the river to understand the world around them. Our approaches, and our knowledge of what lies ahead, are very different—the river is so much more managed now—but when I'm alone, with high water rushing past me, trying to blindly feel my way through an unknown rapid, I still feel like I'm leaning into something unknown, and exploring. I know the river isn't exactly wild, but it feels untamed to me.

PROTECT THE GREEN
RIVER AT ALL COST

After Red Creek, the river bellies out and slows down. The canyon walls drop and slope away, opening into the valley of Browns Park. This is Wild Bunch country, at the border of Colorado and Utah, just below Wyoming. It's isolated enough that Butch Cassidy and the Sundance Kid holed up here in the late 1800s, skipping between states when their luck ran out in one or another. They traded horses with the few settlers here, who were usually happy to oblige. They were all operating within their own set of rules. It was lawless then, and it still feels untamed, especially after the fishing boats trail off and I'm left alone in the echoey badlands, the ghost shadows of trout below me. I stop for lunch near the abandoned Jarvie Ranch, where Scotsman John Jarvie ran a store and ferry crossing until robbers murdered him in 1909.[1] When I put back on the river, the sky has gone to lead, and I can hear the zippered rumbling of thunder in the distance. Upstream, the canyon I came from is dark. Back when I was a raft guide, and afternoon storms would roll down the Arkansas or Eagle River, I would tell customers that the safest place they could be was in the middle of the river, in the bottom of a boat. "We're the lowest thing around," I'd say. But alone, acutely conscious that no one knows where I am, I start to wonder if that was a myth. I start to second-guess the stories I've always told myself about what is good and safe and right.

The more time I spend on the river, and the more I talk to people who operate in and around dams, the more I start to question the assumptions I've made about them, too. I'd come in with the biased view that dams were, for the most part, bad, environmentally destructive, and stupid, and that they should be removed wherever possible. But the more I talk to people who touch the water every day, the more I realize how naïve and unsubtle my thinking has been. A lot of that still holds true, but there's no hard-and-fast rule for what best serves people, the surrounding ecosystem, and the river all at the same time.

People like Kirk stress the importance of storage, and of controlling our water so it can get to people when they need it, because we've become dependent on consistent stream flow. "People are the most important factor" is a refrain I hear from people all along the river, from fish biologists in Pinedale to National Park employees in Moab. They all mean slightly different things, but the heart of the matter is the same: people need water to survive. But to get water to people, and to sustain those constant water supplies, we have to sustain the rivers, and those intentions can be at odds, or at least hard to balance.

The 1900s were the years of the government's concrete dreams, of holding back rivers to help make deserts more habitable for people. By 1979, when the Bureau of Reclamation turned on the power plant at the last large-scale federal hydropower project, New Melones Dam on California's Stanislaus River, most of the rivers that could create viable storage or hydropower had been dammed up.[2]

There are an estimated 90,000 dams in the U.S., ranging in size from inches high to the Hoover Dam's 726 feet.[3] Dams of any size irrevocably change complex river ecosystems. As soon as water backs up behind a dam or diversion, it moves through its former channel differently. Stagnation alters both physical and chemical characteristics. When dams artificially slacken water that used to move, they alter its temperature, its dissolved oxygen content, and the ability of native plants and animals to live in it. Reservoirs, which emit methane, contribute to greenhouse gas emissions. The sweeping canyons of Flaming Gorge are beautiful, but the reservoir is a constructed, changed ecosystem, and, as I'm learning every day, beauty and nature aren't necessarily the same thing.

The 2000s brought a wave of backlash against existing dams and an effort by environmental groups, like American Rivers, to remove structures deemed unnecessary, environmentally destructive, or dangerous. That dam removal movement has worked really well in some places. After the Elwha and Glines Canyon Dams, on Washington's Olympic Peninsula, were removed, in 2012 and 2014 respectively, native salmon populations rebounded almost immediately, and the river's delta stabilized, thanks to increased sediment loads, creating habitat for elk and other animals. In Maine, when the Veazie Dam came down on the Penobscot River in 2013, fisheries came back, boosting the economy in the area. But a lot of the successful dam removals have been on rivers where water storage isn't important, or where power-generating infrastructure is no longer relevant. The cause and effect isn't always as straightforward as those projects have been.

The reservoirs of the Colorado River system, from Fontenelle down to the Morelos Dam on the U.S. border with Mexico, are the largest in relation to the river's stream flow in any watershed in North America.[4] The river is stopped up, slaked off, and used up to the point that it dries up a hundred miles inland from its delta. But if Fontenelle Dam were to disappear tomorrow, Rock Springs would be dry by midsummer. In a human-constructed and constrained system, the positives and negatives aren't in balance. You can't look backward and think about ripping out dams without taking the present into account. People are important. An idealized version of what the river looked like before the dam isn't an accurate or useful goal.

History is hyper-present here, and since I haven't seen another person for hours, I have plenty of time to think about the people who have come through here before me—and not just because I can imagine where Jarvie might have been murdered, and how desperate those robbers might have been, traveling across this empty landscape. After Powell's party battled their first real rapids in the canyons of Flaming Gorge, the openness of Browns Park came as a relief. They shot game for fresh food and stretched out in the sun after the deep canyons. Powell, a hard-line former Civil War major, who took his role as a surveyor incredibly seriously, didn't give his men many breaks, which one of his crew members, George Bradley, complained about in his journals, but he eased off on them here. As the water slows down and turns chocolate-colored with sediment, I think about how their bacon was rancid by this point on the river, only a few weeks into their trip, and how their wool layers would have been corroded with sweat and salt. Their sense of what was downstream, and of how the water was coming to them, was shaped only by stories and by what they'd seen so far.

A lot of the anti-dam arguments put forth by environmental groups and scientists are based in hydrology and biology. Dams will stress a watershed almost beyond repair, and they fundamentally change the river's function. Jack Schmidt, a fluvial geomorphologist and professor of watershed sciences at Utah State University, who has spent much of the past four decades working on the Colorado River system, says he thinks it's impossible to fully restore pre-dam conditions. Despite that, he's constantly trying to recreate the past, and to help manage rivers in as natural a state as possible, to maintain ecosystem integrity. The hard part is figuring out what counts as natural and how to preserve multiple aspects of the river at once: to manage for native fish and the fly-fishing economy that's sprung up below the dam at the same time, or for both power generation and stable base flows.

Jack, who is a paddler and a river junkie in addition to being a sediment nerd, has been looking at the Green since 1992, when the superintendent of Dinosaur National Monument asked him to evaluate the impact of the Flaming Gorge Dam on the rest of the river. That's his specialty. Fluvial geomorphology is the study of how a river changes itself physically: how the channels narrow and vary, and how flood patterns—man-made or otherwise—affect the landforms and sandbars along its course.

Those changes are both consistent and mysterious. Allegedly, when Albert Einstein's son Hans declared that he wanted to study fluid dynamics, Einstein asked him why he'd ever want to look at something so hard. Geology adjusts on a granular scale. As rivers flow downstream, they're constantly moving sediment loads downstream, too, in the slow rolling force of geologic history. Like race cars, rivers move faster around the outsides of corners, building up sediment on the insides as they lose centripetal force. Migrating sandbars switch side to side as they move downstream, and they vary over time, alternating cut banks and point bars, eroding from one side of the river and loading the other bank.

Spring floods set the sediment in motion as they flush more powerfully erosive flows downstream. On the Green, below Fontenelle and Flaming Gorge, those floods don't happen like they used to because of the dams—if you have smaller floods, you have less pushing. Even with the dam-created spring peak flows for the fish, the river can never run higher than the 8,600 cfs that Flaming Gorge will release with all the floodgates open. Sediment accumulates in reservoirs, and dams change the way alluvial fans build up: they can stagnate and grow over, or they can wash downstream. That's what Jack is trying to understand. In Browns Park, where I am now, the river is between 10 and 25 percent narrower than it was at the beginning of the twentieth century.[5] He thinks that's because of the lack of early-season floods and the resulting ecological changes. Non-native vegetation, like tamarisk, which thrives in the alkaline soil here and has long, many-tentacled root balls that hold riverbanks together, has changed the channel size. Those physical forces then trigger biological outcomes—they break down habitats and change what kinds of species can live where. "Scientifically, we have to worry about the flow regime, how the timing and the magnitude of the flow regime triggers natural ecosystem processes," Jack says.

Those impacts carry far downstream, and they're exacerbated by drought, climate change—which comes with variable runoff and increasing temperatures—and those non-native fish and plants that have spread across the river's system. Jack says that, all things considered, the Green has the healthiest river ecosystem of any part of the Colorado River Basin, in part because the untouched Yampa River, which flows in just south of here, adds so much to its flow, but it's still highly changed. Nothing is untouched anymore. When a river's flow is reduced or altered, which is happening across the West, it's less dynamic, degraded, simplified. And that means it's less naturally resilient.

In the farms, flats, and canyons of Browns Park, the rock layers transition back and forth between chossy white eolian sandstone and the striated red of the Proterozoic Uinta Mountain Group, some of the oldest rock on Earth. The only helpful piece of scouting information I'd been given was that the river was rushing high enough that I probably wouldn't make it under the Taylor Flats Bridge, one of the only places to cross the river in the valley, without decapitating myself. Dropping toward it, I can tell that the guy at the fly-fishing shop was right—there is only a sliver of light between the water and the rebar of the bridge. I pull my boat over and grab my dry bag and paddle, starting to shuttle gear.

There's a white SUV on the bridge, which is the only sign of humanity I've seen in a while. As I schlep gear to the banks on the far side, it drives toward me and pulls up close. A large man with an ACE bandage on his right wrist and a turquoise cuff on his left sticks most of his upper body out the window, beckoning me over. His passenger seat is filled with burger wrappers and half-opened maps. "Look at these rocks!" he says, launching into a story about his time in the canyon.

When he asks if I'm alone, I freeze and fib. My friends are just up-stream, I say. I have a knife and a Pepto-Bismol-pink can of Mace in my dry bag, but one hasn't been sharpened since 2001 and the other has never been used. I think that he's probably just lonely, but he trails on a little too long, not picking up on my cues that I'm ready to leave the conversation. I shift from foot to foot, uncomfortable and ready to move downstream, but I don't want him to think that I'm skittish. I make up an excuse and break off when I can, heading back to my boat.

There's a stigma associated with women alone in the wilderness, and a long-running narrative that women go into the outdoors alone only when they're pushed, when they're running from bad decisions, or when something traumatic has happened to them. Adventure for the sake of it is a part of a long-held western narrative, but it only seems to stick to men. I don't like the idea that you have to be exceptionally brave, or exceptionally desperate, to go into the backcountry. I don't think that's true. Not from my own experience before this, or from what I see my friends doing, or from my time on the river here. But sometimes I still question why I'm here, what I wanted and what I was trying to prove. Whether it was the adventure or the aloneness that felt important, and how they're tied together, in my own head, and in the eyes of outsiders. I still feel vulnerable, and alone. Without anyone else to reason with, I'm not sure when I'm being stupid and when I'm being brave. The shiver of that vulnerability hits me, and I paddle downstream as fast as I can, looking back over my shoulder, even though I know it's obvious where I'm going. At this point it's not the river that scares me, it's what people think when they know I'm alone. I think there's value in beating through that stigma and the silence, to mark some little truth against it, so I make myself push on, not letting the fear of lightning strikes or leering strangers stop me. Isolation is tricky, because it's easy to lose a sense of what's really a threat.

But I like being able to make up my own mind, too. Out here alone, where I have space to focus on the river, the interconnectedness feels vivid and impossible not to think about—the way the water slowly changes from clear to murky below the dam, or the intakes pump water up to wells in surprising places. It changes the way I think about the river as a whole, because I'm realizing how much I was wrong, or narrow-minded, about.

Sometimes, to show the big picture, it's helpful to focus on something small. To gauge how dam operations impact everything below the power plants, scientists have been looking at one of the bases of the food chain, the insects. Because the bugs are close to the bottom of the riparian food web, they're bioindicators for the whole system, indicative of the health of the river.

On the Green, USGS research scientist Anya Metcalfe has been looking at insect diversity in the river, and she's started a citizen science campaign to collect data. She sends boaters like me out with a bug-trapping kit. I carry a rectangular metal ammo can painted sky blue and filled with bottles of ethanol, a light, and a paperback book-sized tray. At night, after I make dinner, I set the tray on the riverbank, pour one of the ethanol bottles out into it, and leave the light shining on it for an hour. When I come back, it's often filled with the carapaces of dinosaur-like bugs, a sample of what's living in the section of river I'm on. "It's a proxy to look at the river," Anya says. "It's a really powerful tool to understand the availability, what bugs are there, and how many are there."

Anya is trying to get a sense, over time and space, of how many bugs are in the river because that can show how many fish can survive there. She spends a lot of time on the river—I end up running into her on my very last day on the water—but even if she were out every day, it would be impossible to track every stretch of water at every time of year. So she's teamed up with government agencies, guiding companies, and private boaters, who take the kits with them and bring back samples. I slowly build up a stash of used bottles, filled with dead, dark bodies suspended in liquid. She says that on the Colorado River, where they started the experiment, they had data from almost every day last year. The goal of the project is twofold. She's gathering a vast map of data, but she's also hoping that the boaters who capture the bugs start thinking about what it means, and why they might not find any insects in one place, but a lot somewhere else. She thinks it's a line in to caring about the river, and a way to give people a stake.

Anya is spearheading bug tracking on the Green, but she and other researchers in her lab at the Grand Canyon Monitoring and Research Center have been tracking insects in the Grand Canyon, below Glen Canyon Dam, since 2012, and they've found that the impacts of the dam are significant.[6] "Something that became immediately obvious is that there's not a lot of diversity. Midges, blackflies, micro caddisflies—that's all we've been finding," she says. "With classical ecology in the river continuum you're supposed to get more diversity as you go downstream, but the recovery rate is really slow. 150 miles from Glen Canyon we still have low diversity."

They're not seeing many types of bugs even though algae, the insects' main food source, is healthy. They're trying to figure out why, because without bugs, the fish are hungry and struggling. They think that peaking the dam's releases based on hydropower needs might be washing bug larvae downstream, and that temperature might play a role, too, so they're trying to figure out what dam operators can do to create better conditions for bugs and fish, how they can use man-made tools to mimic natural cycles, or mess up the river less. The water coming out of Glen Canyon Dam, which comes from the bottom of the reservoir, is incredibly cold, and the researchers think that's rough on the native flora, which was historically used to warm river temperatures. At Flaming Gorge Dam they've installed what's called a selective withdrawal system, which lets them pull water from whichever layer of the reservoir they want, adjusting the temperature of what's released downstream. They think that helps bug viability downriver. "Immediately below Flaming Gorge looks a lot like the Grand Canyon, but fifteen to twenty miles downstream you start to get the foodstuff bugs like stone flies and caddisflies," Anya says. "They're classic indicators of healthy, happy ecosystems."

Anya says she's excited by the idea that dam managers are thinking about bugs and are working within the framework of the system to try to manage for them. "We can't restore the river because it would require taking out the dam, but we can understand how the dam impacts ecosystems, then try to understand how to best manage it," she says. "You can't always manage for restoration, but you can try to manage for a healthy ecosystem."

Healthy is a tricky, subjective goal. Jack believes that for the future of healthy rivers, and to hold onto some modicum of pristineness, we should try to keep the Green as wild as we can. The dams have broken the rivers, that's inarguable, but they're there. So now, he says, it's important to try to manage the rivers in as natural a way as possible and to gather as much data as possible to try to figure out what natural looks like. It's a combination of looking realistically at the ecosystem and working smartly within the bounds of what we need to exist.

If Jack could change the way water is managed, he says, he would deemphasize the Colorado River Compact and put more value on environmental flows. "I'd make the Green as wild a flow regime as it could be. I'd do everything we could to protect it. I'd say no to every new proposal to divert water. I'd say that the highest and best use of the Green is to let the water flow downstream and let someone use it downstream. I would protect the Green River at all cost."

THE MAP OF WHAT'S NEXT

Partway through Browns Park, the blood-colored Uinta Mountain Group rocks break back into the rolling brown valleys. "A spur of red mountain stretches across the river, which cuts a canyon through it," Powell wrote on June 4, 1869, after he paddled into the ruddy gorge. This geologic injection, which he named Swallow Canyon, is achingly beautiful, with the river pushing quietly through the red and the green. The water is a color between chocolate and slate, and it absorbs light. In the depth of the canyon I stop moving, rest my paddle across the skirt of my boat, and listen, letting the river spin me, catching the whirlpools of eddy edges. I don't think I've ever been this alone, and I get a thrill from the idea that I have the canyon to myself. That no one knows where I am. I can hear the sizzle of water off the end of my paddle and the thwap of duck wings as they skim the water, taking flight. I've never regretted my lack of knowledge of birds until now. A great blue heron fights gravity, rising slowly.

That thrill turns to frustration when I realize I've accidentally floated past the spot where I'd intended to camp, and I have no one to blame but myself. Without anywhere to stop, the canyon suddenly makes me feel trapped and anxious, but as the light starts to peter out the gorge opens up again, and I find a slot in the tamarisk where I can pitch my tent. I set up camp and make curry from a packet, reveling, again, in the quiet. I have Terry Tempest Williams to read and the ghosts of Butch Cassidy and the Sundance Kid for company, along with a jackrabbit who keeps circling closer. I can hear the click of the pivot sprinklers on a distant hayfield. I think all the wrong things are beautiful: invasive cheatgrass, the glint of sprinklers firing in late-day light, the glossy introduced rainbow trout. I realize it's hard for me to tell what's native and natural, what's been altered by people, and what counts as history.

The Green, and the reservoirs behind the Flaming Gorge and Fontenelle Dams, are part of the interlocking storage system for the whole Colorado River Basin. Not far past the Green's confluence with the Colorado, the combined river runs into Lake Powell. Past the gates of the Glen Canyon Dam, which forms Powell, the river then runs through the Grand Canyon into Lake Mead, the largest reservoir in the U.S., formed by the Hoover Dam. There's about 60 million acre-feet of storage capacity in the Colorado River Basin, which is four times the average annual flow in the river. Mead and Powell account for 50 million acre-feet of that capacity. Together, they're the river's savings account. Powell, which is right above Lees Ferry, the dividing line between the basins, was built to ensure that the Upper Basin could meet its compact obligations; Mead supplies the Lower Basin. Or at least that's how it's supposed to be. In the past, that quadrupled storage has been able to balance supply and demand in dry years, but now demand has risen, and the reservoirs are getting low. Stored water is even more important in drought years, but drought depletes it. The more we need it, the less we have. Between evaporation, reduced inflow, and increased use, the West is sucking itself dry. The reservoirs are intended to form a buffer, but the system breaks down when there isn't much water coming in.

It's been dry in the Colorado River Basin for all of this century. Years ping-pong—some are above average, some well below—but they've been trending steadily dry. In 2017, climate scientists Brad Udall and Jonathan Overpeck found that inflow to the Colorado River was reduced by an average of 19.4 percent between 2000 and 2014, and that one-third or more of that decline was due to global warming. In addition to low inflows due to lack of precipitation, evaporation from the reservoirs removes over 10 percent of the river's flow, a huge proportion.[1] The major reservoirs haven't been as low as they are now since they first started filling up in the 1930s, and for the first time, that's starting to affect water rights. It could change how people can use water in the basin and how much water they might have access to. Between 2001 and 2015, Lake Mead's surface dropped from 1,196 to 1,075 feet above sea level, an elevation that triggers cutoffs and shortages in the Lower Basin. According to the 2007 interim guidelines for the basin—one of the first changes to the compact that addressed the possibility of less water in the river—when Mead hits 1,075, the Bureau of Reclamation can reduce water deliveries to Arizona and Nevada, giving them less than their allotted compact water.[2] Low levels in the lake also mean that water might not reach the hydropower turbines of the Hoover Dam, which send power to seven states. And that affects more than just people who want to run their air conditioning in Vegas. Hydropower bills provide funds for restoration and remediation projects, which support the endangered fish, so, in a tangled way, the fish depend on the turbines.

In the beginning of the twenty-first century, as water managers watched reservoirs drop toward their minimum power pool—the lowest level at which it's possible to run the turbines—the need for adaptation became clear. That's why the Department of the Interior came up with those interim guidelines for how to operate the reservoirs along the river in times of water shortage. The guidelines were a large-scale attempt at planning for ensuing drought, so that a framework would be in place to ameliorate future fights over water. That kind of interstate coordination hadn't happened previously, and the interim guidelines were hard fought while they were being written and are unpopular in a lot of places. But something needed to happen.

As Lake Powell and Lake Mead drop, complicated questions surface about levels of risk. For instance, if those two major reservoirs run dry, where else can water be stored? Wyoming is banking on building small-scale storage up high in the basin, where cooler temperatures lead to less evaporation, but those Wyoming reservoirs aren't likely to add up to enough storage space to offset losses in the major reservoirs. The big benefit of Lake Powell and Lake Mead is that they're incredibly large. A vast amount of the last century's urban growth—cities like Los Angeles, with its huge economy—was built based on the ability to import water from those reservoirs. Cities have grown up assuming that their residents will have water at their fingertips when they turn on a tap, and no reasonable person wants to cut off water to a city. For the river system to be sustainable in as many ways as possible, it's not just the amount of water that's stored and shared that will need to change, it's also how the water comes downstream and when, and both water managers and scientists are trying to figure that out.

"The fundamental issue is trying to separate the impacts that the dam itself is having from the issues related with management," Ted Kennedy, Anya's boss at the Grand Canyon Monitoring and Research Center, says about Flaming Gorge. "Nobody talks about taking the dam down, at least not seriously, but how do you manage ecosystem restoration around and including the dams? How do we disentangle these different stressors and isolate the ones that are related to things that we can change, like the pattern of water delivery, and still allow for hydropower? The big thing in science, when you have multiple things that have changed, is that you can't run a bunch of experiments next to each other, like flow management with this backdrop of sediment and flow regimes. It gets really tough when you try to pull the threads."

When you're trying to manage for nearly a dozen factors, ranging from insect diversity to power generation, there is no past success story to emulate. It's unrealistic to think that the river will ever look like it did before dams were built, so managers are trying to manage for a new equilibrium through what they call adaptive management. The best way to do that, and to make as many different groups as happy as possible, is to get them to talk to one another and to try to build consensus. No one likes top-down government mandates, not even the government employees, so they're trying to figure out collaborative management plans for the dams and reservoirs.

In the flats of Browns Park, I wake up early, as soon as the sun turns my tent into a sauna. I boil water for packets of coffee and oatmeal, remind myself to roll back my shoulders and stretch, and repack my sleeping bag and stove into my dry bag. I'm starting to hit a rhythm. Each ensuing morning that I load the boat, I'm less nervous that I'll have inadvertently scratched a hole in it and that it will no longer hold air. I know, on some level, that the routine is tenuous, it depends on nothing going wrong, and I'm not entirely sure what to do if it does. I have a roll of duct tape, an emergency locating beacon, and some heavy painkillers in my emergency kit, but there's no way in hell I want to use the last two. Part of being out here is that suspension of disbelief, convincing myself that if anything falls apart I'll be able to deal with it, even if it's a problem I've never seen before.

As I move downstream through Browns Park, the river is browner and cloudier, the hills are lower, and everything feels more relaxed. Micro changes are constant, even if I can't feel them. These canyons were carved by the river cutting through rock, slowly eroding it. Every grain of sediment is a rasp, scraping layers off the cliff bands. The river is eroding the human-made mountains, too. Dams have a limited lifespan, so over the next few decades, the huge ones built in the last century will start to break down, and they'll need refurbishment and maintenance to avoid disaster. The government will have to decide which ones to hold onto and sustain. By 2020, 70 percent of the dams in the United States will be more than 50 years old,[3] and of those dams, 17 percent are considered "high hazard." If they break, there will be destruction in their flood paths. The dam removal movement has sprung up both because of that and because of the inherent damage dams do to rivers. Groups like American Rivers have been working to take out the ones that are dangerous, or were built for reasons that are no longer relevant, or were unnecessary from the get go. There are plenty of dams in the country that were built to raise funds through taxes, or for agriculture that never panned out.

That historical context explains a lot about both the benefits of trying to control water and the destruction that results. I try to assume that people were always trying to do the right thing at the time. "If people want to know about the Green, the first thing that they should do is start from the start and read a little bit about Powell's trip," Steve Wolff, who works at the Wyoming State Engineer's Office, told me, before I took the trip. "You can't have a romantic view that's not informed by history. A lot of people want to dynamite every dam, and they don't understand what would happen. I think Powell was a visionary in defining how you could do development in the West. The history of the

West is the history of the Colorado River, whether you're in Denver or Phoenix or St. George, Utah. If you value the kinds of opportunities we have in the West you value the Colorado River. To continue to live in the desert we have to be pragmatic and careful in how we manage the resource. We can work with each other, we can be scientific and solution minded, and respectful of each other, but if we don't plan very carefully there could be incredible effects."

So here, hopefully, adaptive management, and learning how to operate the dams in imitation of what the river would be like without them, will drive water storage. It will take micro changes, as well as a search for compromises that work for as many people as possible. No facet of dam management is ever unilaterally popular. As Jack Schmidt says, it's easy to control for one factor, but almost impossible to manage for multiple factors.

I paddle back out into the turgid river, trying to get to the end of Browns Park by the time it gets dark. I'm looking for markers that line up with my map, the rib of a vague topographic contour line, or a significant turn, but the river ahead and behind looks the same. I'm still frustrated by not knowing exactly where I am, antsy when I don't feel completely in control. I try to settle in, try not to get fidgety and anxious. When I'd peeled out from the boat ramp below Flaming Gorge, I hadn't known how much I'd be juggling the pluses and minuses of every piece of water use in my mind.

FISH

9,080 CFS

LARVAL TRIGGERS

Aside from the A section right below Flaming Gorge Dam, which is full of trout and fishermen, most of the river I've floated so far has been empty. I've gone days without seeing another person, much less a boat, but that changes below Browns Park. Some of the desert canyons downstream are popular enough that they require hard-to-get permits. I have time to kill because of the way my permit dates line up, and I spend it off the river in Vernal, Utah, which feels like a living Venn diagram of competing water uses. Highway 40, a string of fast-food joints and long-term-rental hotels geared toward oil and gas contractors, runs through the center of the city, and the downtown strip looks like it's been crumbling since most of those contracts ran out. White oil-field trucks are parked next to rafting company buses, which drag trailers stacked with boats. Dinosaur National Monument is just east of town, and hay farms, which have been around since the Mormons settled the area in the 1860s, dot the margins. Unemployment is currently twice the national average because drilling in the Uinta Basin has tapered off sharply due to low oil prices.

Vernal is also an outpost of one of the most controversial water users on the river, the Upper Colorado River Endangered Fish Recovery Program, which is responsible for protecting four endangered fish species.

In the back room of the U.S. Fish and Wildlife Service office, fish biologist Tildon Jones pulls a three-inch-long glass bottle out of a cardboard box and holds it up to the light. He spins it, swirling razorback sucker larvae that look more like fingernail clippings than fish. They're just silvery white slivers with dark indigo eyes, but these endangered fish spawn set off the chain reaction that determines when Flaming Gorge Dam starts peaking its flows. This time of year, in the late spring spawning season, Tildon's job is to look for the first signs of larvae. His team sets light traps in the backwater channels where suckers spawn, trying to catch the ghostly scrim of future fish. When they find the first larvae of the season, they call Heather Patno at the Bureau of Reclamation and tell her to start releasing water from Flaming Gorge. Then the dam opens its floodgates all the way, sending 8,600 cfs downstream for the sole purpose of helping razorback suckers flush into their spawning habitat. The tiny translucent fish, along with three other endangered native species—humpback chub, bonytail, and Colorado pikeminnow—are the heart of an adaptive management program that affects water use all along the river.

Those native fish weren't always so important. For most of the twentieth century they were considered trash fish: unattractive, unappetizing, and no fun to fish for. By the time the Green was dammed up their populations had plummeted. The dams turned the warm, silty water cold and clear, changing the food web and altering their habitats. The fish were also targeted directly. In 1962, as Flaming Gorge Reservoir filled up, the U.S. Fish and Wildlife Service decided to poison the river with the piscicide rotenone.[1] The plan was to get rid of the native fish population to make room for rainbow trout, a sport fish that was considered more valuable because trout fishing brought in tourists and their dollars. Fish and Wildlife dropped rotenone into 430 miles of the river starting right below the dam before they released the trout.

By the time the Endangered Species Act went into effect in 1973, some of those native species were close to extinction. The poisoning was just part of it. Dams had destroyed their spawning grounds, and predatory non-native fish, like smallmouth bass, had swum upstream from stocked ponds, outcompeting the native fish for their homes and eating their young. In the decade after the rotenone poisoning, the environmental movement had changed the tenor of the conversation around species diversity. Instead of being seen as a nuisance, the ungainly native fish were starting to gain recognition as a crucial piece of the ecosystem and an indicator of the health of the river.

The U.S. Fish and Wildlife Service added the humpback chub, bonytail, and Colorado pikeminnow to the endangered species list in the early 1970s, which—because the ESA is federal law—guaranteed them federally reserved water rights, which carry precedent over state water rights. That decision didn't sit well with other water users. The Colorado River Water Conservation District sued Fish and Wildlife on the grounds that the listing encroached on the district's water rights. Nearly a decade of litigation followed. In 1988, Fish and Wildlife, the Upper Basin states, and the Bureau of Reclamation agreed on a multi-party recovery program for the three fish. When razorback suckers were declared endangered in 1991, they were added to the program.

Fish and Wildlife was charged with rebuilding the native fish populations, so their biologists started working on plans to bring back the fish within the bounds of both the dam-created ecosystem and the power needs downstream. "The last remaining habitat for those four endangered fish is from Flaming Gorge and Yampa Canyon to the confluence with the Colorado," Tildon says. "The goal is to recover the four species while still allowing for the development of water according to the compact."

Tildon says they decided that the best way to protect the fish was to return the river to as natural a habitat as possible and bring native fish populations back to the river. They're doing it by mandating flows, restoring wetlands, stocking native fish, and getting rid of non-native species. None of that is easy or straightforward. The program didn't start until the populations were diminished and their habitats were partially destroyed, so their baseline is blurry. By the 1980s there were no more reproducing populations of bonytail in the wild, and pikeminnow spawn were found only in two isolated pockets of the river. Tildon says they're constantly guessing at how best to bring them all back.

When the recovery program began, no one was sure which biological variables were affecting the fish the most, or which species' reproduction was triggered by what facet of the water cycles, so they tried to control one variable at a time. There's a single spawning period each year, so scientists only have one shot at collecting data on fish behavior, and it takes multiple years to see any meaningful patterns.

Sometimes they got unsubtle clues from the universe. In 2011, the third biggest water year since the dam was built, Flaming Gorge released a high volume of water over a longer than usual period to keep the reservoir from overtopping the dam. The water wasn't released for fish—it was dumped to maintain the dam's integrity in the face of flooding—but downstream wetlands were connected to the river for longer than usual, and scientists found razorbacks in those wetlands. Suddenly the researchers knew more about the fish's natural habits. They learned that razorback suckers spawn in flooded back channels, and that the fish need the flush of spring floods to get into those spawning habitats—their spawning is timed to the peak of the hydrograph. Tildon says that razorbacks lay eggs exactly when they'll hatch with the peak flow, and it's eerie how closely the fish seem to be able to predict the peak. "They're intimately adapted to know, they have this sense of when it's going to happen, and their timing is impeccable," he says.

When the dam releases were based solely on power generation, flows never peaked, and the suckers couldn't spawn. "We started the larval trigger study in 2012. Before that we had 20 years of data where the release was mismatched with hatch," Tildon says. "Wetlands were flooding, but as the water dropped the wetlands drained. The fish can't swim against the current, so they were getting stranded. Now, we raise the river and it flushes them into the wetlands. The larvae are getting into where we want them, then after that it's a matter of making the wetlands suitable for them to survive."

The recovery program started asking the Bureau of Reclamation to open the Flaming Gorge bypasses as soon as they saw the first spawn downstream, even if it meant ignoring power needs. Because the Endangered Species Act is a federal law, the bureau had to comply, and those larva-triggered fish flows are now one of the major determinants of how water is released from the dam.

The Vernal headquarters of Fish and Wildlife is a mullety block of beige government cubicles with an unruly garage full of gear behind it.

The back room of the office is stacked with straps and boat parts. Snapshots of trophy fish and sketches of larvae are tacked on the walls. Tildon pulls one of the light traps off a shelf to show me. It looks like an inverted old-school camping lantern. He says they stage the traps at the flooded mouths of tributaries, the beginning of the off-channel habitats. When they set a trap in the river, they put a glow stick in the bottom. The larvae swim in, attracted to the light, and get stuck. This year they found the first signs of razorbacks in Jensen, just downstream of Dinosaur National Monument.

The larval trigger has helped the razorback sucker to spawn, but that's just one segment of one tiny fish's fragile life cycle. Tildon says that trying to manage for four fish, which don't depend on the same flood regime, can be a nightmarish guessing game. They've been able to rebuild populations in hatcheries, but they're still struggling to figure out how the fish live wild in rivers and what kinds of conditions help them thrive. They think stable base flows are more important for the pikeminnow, so ideally they want the river to peak high in the spring, then sustain a base flow through the summer, but it's hard to ask the Bureau of Reclamation for both a flood release in the spring and a steady base flow in the summer—both politically, because it affects other users, and because of the finite water supply.

Seasonal workers, most of them skinny college-aged kids, are rigging gray rafts in the garage. They are packing coolers and ammo cans full of supplies for a week on the river. They're planning to do population estimates for the pikeminnow and razorback sucker over two sections of the Yampa and the White River, which runs into the Green south of here. To do that, they'll electroshock the river, sending a current that stuns the fish through the water. The fish float to the surface, where the biologists can count and catalogue them. They've been putting PIT tags, individually coded thirteen-digit passive integrated transponders, into the fish they find on the river, as well as the ones they release from the hatchery. The shock brings up non-native fish, too, which they're trying to eradicate. Later in the summer they'll spend a lot of time removing those fish from the river. On this trip, they'll kill all the non-natives that come up. Sometimes they eat invasive walleye for every meal.

Tildon is lanky and southern talking, smart and thoughtful, and he's hyperconscious that what he's doing is politically hard. A big part of his job is explaining to people along the river why the recovery program is important. Many people around here have a baseline distrust of both federal government and environmentalists, because there's a long-held dichotomy that pits environmental regulation against economic drivers. The fish have become a symbol for many of the imbalances within the river system, particularly among ranchers and other water users who feel that the fish unfairly take priority over human livelihoods. Tildon spends a lot of time getting screamed at.

The native fish are indicator species, and their vitality indicates the shape of the whole river system. If they're healthy, it broadly shows that the ecosystem is resilient, which is why Fish and Wildlife has put so much stock in saving them. In the sphere of adaptive management, tying the dam-controlled river to natural flood patterns can be a way to promote a much healthier habitat overall, but it frustrates people whose lives depend on the river in different ways.

There's a fish tank in the Fish and Wildlife office that holds a few of the native species. Inside, the endangered fish—suckers with their tiny, priggish mouths, top-heavy humpback chub—look like dinosaurs. None of them are attractive; they're the opposite of the sleek, colorful rainbow trout that have been stocked below the dam, but they're adapted to thrive in the variable, sediment-filled waters of the Green. They live for up to 60 years, so the older ones have had to be flexible in the face of drought cycles. "These fish have been in this highly dynamic system for a few million years, they've made it through changing conditions," Tildon says. "That's the only thing you can say about the Colorado River Basin, it's always in flux."

The recovery program is trying to figure out how big those fluxes can be: What are the variations that push the fish too far to survive, and what have humans added to or taken from their habitats that has made survival untenable? Once scientists know how the fish live in the river, they can get to work protecting it. But it's hard to map out a life cycle for a fish whose record is spotty or nonexistent. And there are so many variables. "Bonytail almost went extinct before we could figure out how they acted in the wild," Tildon says. "All the bonytail in the world were caught in a single lake in the Lower Basin."

Their habitats are at risk from more than just dam-released flows. The Colorado pikeminnow, which can be six feet long and weigh eighty pounds, migrate hundreds of miles down the river to spawn. There are only two known spawning bars in the river system. One is in Yampa Canyon, in the last wild stretch of the Green's basin; the other is farther downstream in Desolation Canyon. They're both on cobble bars, where the river flows over an amalgam of baseball-sized rocks, which oxygenate the water and slow it down a little, giving the fish places to lay their eggs. Tildon says they don't know exactly why the fish use only those two places, whether it's some kind of ancient homing instinct or particular conditions. He says it seems as if there would be other places that create similar atmospheres, but the fish are tied to their historical breeding grounds. He worries that those fragile habitats will be destroyed by human impacts before the recovery program can unspin all the mysteries.

Vernal is the heart of the Uinta Basin, home to some of the most significant oil and gas lands in the country. There's a web of pipelines crossing the rivers and streams in the watershed, sending fuel from wells to refineries. And those pipelines have accidents. They're rare, but oil spills or other chemical leaks can devastate the river. One of the pipelines that runs through the basin is the Chevron line, which spilled at Red Butte Creek, along the Duchesne tributary, in 2010. The spill resulted in seven years of drawn-out cleanup and unresolved litigation. Chevron had a pipe burst in the Great Salt Lake recently, too. For the number of miles of pipe, the number of spills is relatively small, but the damage is significant where they happen. And it's not just oil. Salt from the fluid used in hydraulic fracturing is particularly harmful because it dissolves in water, which makes it omnipresent and hard to remove.

There's an oil pipeline under the Green at Ouray National Wildlife Refuge, just upstream of Desolation and Gray Canyons, home to one of those two pikeminnow spawning bars. Tildon helped prepare the Environmental Protection Agency's emergency response plan for the Green River, which looked at what the worst possible scenario might be if that pipeline burst. There's no access to those canyons beyond a single road at a place called Sand Wash, where boaters put in, so if there were to be an oil spill in the refuge, and it wasn't stopped before Sand Wash, where the canyons close up, it couldn't be stopped for nearly a hundred miles. He says that it's almost impossible to predict when a pipe will break or a valve will fail. Accidents happen all the time, and the industry isn't known for its adherence to environmental regulations. He's seen wells that were drilled illegally in the floodplain get washed out, and a wellhead and a containment pond break in tandem, right before a flash flood, spilling fracking fluid into the river. Any of those accidents could decimate the already fragile endangered fish populations. "We can stock more razorback, but I don't have pikeminnows in a hatchery and it takes six or seven years to get them to the point where they could be released," he says. "They have them in New Mexico, as what they call refuge population, in case something happens, but I don't know how well they'd do in the wild."

The endangered fish have become a thorn in the side of the energy and agriculture industries, and they make power generation and dam management more complicated. They've become shorthand for heavy-handed government oversight. But they're also holding the river close to what it used to be.

HUMANS ARE A SPECIES, TOO

Life in Vernal ticks by hot and slow. I eat gas-station burritos, drink weak Utah beer, and run the mountain-bike trails outside of town at dusk, when it gets cool enough to move in the desert. I camp on chalky flats of BLM land alongside the river and spend a lot of time in the air-conditioned public library and recreation center, which both seem unnecessarily big and glossy for a town this scratchy. The same was true in Pinedale, and in other energy towns I'd been through, where the bones of both oil's boom and its bust are visible. I cruise town, trying to talk to folks at both the drilling companies and the rafting out-fitters. I have more luck with the boaters, who are happy to talk about why the river is important to them. Plus, my markers are clear the second I roll in: my feet are Chaco-tanned and I'm driving my lived-in station wagon crammed with all my gear. When I stop to talk to Bruce Lavoie, who runs the rafting company O.A.R.S., he tells me there's a short-notice public meeting of the Flaming Gorge Working Group later in the week that might be worth checking out.

It's a stifling, airless Thursday night, and the sun is still high when I pull up between a pair of trucks in the Uintah Conference Center park-ing lot. The building is beautiful, newish, and devoid of personality, another artifact of the energy industry's financial clout. It was finished in 2015, just before the most recent dip in oil prices.

The Working Group, the entity that informs the public about plans for the reservoir, usually meets three times a year, twice in April to plan flows and again in August to talk about what happened in the previous season. This is an unusual meeting, right after the peak. The Bureau of Reclamation called it to explain the ways it changed the flow plan in light of the unexpected weather. After a winter of slightly below average precipitation, spring came on cold and wet, and it stayed that way into the early summer. The rush of precipitation didn't line up with the bureau's forecasts, and flows into the reservoir jacked up unexpectedly. To keep the reservoir from overtopping the dam, the bureau's engineers let out a higher volume of water than predicted, starting after they spotted the first endangered fish spawn, and kept it coming. The water flooded farms on the riverbanks and remained there, destroying cash crops. The commercial fly-fishing guides who run the A, B, and C sections below the dam lost revenue because high-dollar fishing customers cancel their trips when the river is too high.

Now all these people are crammed into a conference room on a hot, stale night, ready to air their grievances. They're government employees, farmers, and fishing guides, mainly, and they're sitting in pockets of like-minded folks. I try to pick them out by their outfits: the representatives of the Green River Outfitters and Guides Association in their khaki quick-dry and croakies; Tildon's boss, Tom Chart, the director of the Upper Colorado River Endangered Fish Recovery Program, in an insignia-bearing government polo shirt.

I sit down behind a row of ranchers, Brent and Kalynn Sheffer, and their son and daughter-in-law, whose family has been ranching on the Green for five generations. I eavesdrop before the meeting starts. Everyone is talking about how hard it's been, how the unplanned flows have slaughtered their business, how they feel ignored and disrespected by the government agencies, whose representatives have no idea what it's like to live dependent on the river. They're echoing the complaints that Tildon told me about: they don't understand why the federal government is focusing on inedible fish, but not farmers growing food.

The meeting stays civil through the first presentation, NOAA hydrologist Ashley Nielson's explanation of the forecast and how the heavy spring precipitation spiked the inflow to more than 400 percent of average, swelling the reservoir. She says the forecasters hadn't seen the uptick coming, so the long-range release plan hadn't taken it into account. "This is not what makes a forecaster comfortable, this isn't what we like to do," she says.

Then her NOAA colleague Aldis Strautins gets up to explain flood stages, and how they predicted when and where the river would run over its banks. Kalynn Sheffer raises her hand and asks, "Do you take into account soil saturation?" It seems like an innocuous question to me, but then her daughter-in-law starts murmuring under her breath about flooding pastures, and the heat in the room rises. Their fields flooded, Kalynn says, because those high flows from Flaming Gorge coincided with the peak of the Yampa, which runs into the Green upstream of them, in Dinosaur National Monument. Because the bureau kept pumping water out of the dam, the river came up and stayed up, which had cascading consequences. One older man says he lost about a hundred acres of alfalfa at $1,000 an acre. A younger man in the corner starts yelling about how structures on his property were inundated and destroyed and trees were beat up. How habitats for other animals, like hawks, were flooded and washed away.

The wave of unhappiness comes in a way I hadn't expected. I'd thought that more water would be good, that the dam would bring welcome stability, but the peaking flows that don't necessarily coincide with natural releases change the system. Brent says they've been flooded more since the dam went in. The Sheffers' daughter-in-law breaks out a camera and starts filming.

It simmers a bit, then Tom Chart gets up to explain how the fish program is structured, why they're pushing so much water downstream at certain times, how they want the peak of the Green to meet the rising flood of the Yampa, even though they can control only their half, and why the larva-triggered releases take precedence over other water uses, like agriculture-focused flows. He says the fish recovery program helps the Colorado River Compact states make sure their apportionment is environmentally friendly. He talks through the framework of historical precedent, the legislation, and all the time and money that goes into it. They spend $6 million a year—approximately $300 million over the life of the fish recovery program—that they get mainly from hydropower revenue and farm credits. Facts and feelings collide, even as Tom is trying to line out the logic. The alfalfa farmer asks, "Why do we value trash fish over these people's livelihood? You government workers get to go home and not deal with losing $100,000."

Brent Sheffer, who knows Tom from helping with off-season fish counts in the past, speaks up. "Tom, you know I love the fish, but . . ." He pauses. "I'm all for saving the fish, I'm all for saving the hawk, I'm all for saving my hayfield." When you're living close to survival, like the fish, like the ranchers, it's hard to believe in practices that hit your bottom line. But when he says that it cracks open a gap of understanding.

"I know there's a lot of rub between what we're trying to do and what you're trying to do living along the river," Tom says. "If this program is going to be successful the onus is going to be on the Green River, it's going to be the telltale. Every aspect of the hydrograph is doing something critical for the fish, but the floodplain is also where people live. It's critical habitat, but I realize it's also what's giving you trouble." He says he's been working with native fish for more than twenty-five years because he thinks that if they can preserve the fish populations, they can hold onto the fragile balance of a healthy river. They need clean water, natural flows, and a diverse biosphere, and we do too. It's about the fish, but also about the biological processes that sustain the fish. "There's a human species," the woman in front of me says, under her breath, to Tom's comment about saving the fish species. "But we're not endangered," her husband says.

The fly-fishing guides, whose livelihood depends on the kinds of cold, steady flows that sustain non-native trout, not native suckers, start chiming in about how the unplanned high flows are hitting them, too. They say their bodies are shattered from rowing against the high water, and that they're having to lay people off because so few people want to come fishing when the river is flooding. One of the guides says he lost $24,000 a week during two weeks of high flows. They feel trapped by the federal programs, which they say rarely take their requests into account, even though their economy is dependent on the river.

Most of the fishing operations in the area are based in the town of Dutch John, which was built by the Bureau of Reclamation in 1957 to house workers for the dam.[1] It was constructed in the dam's shadow after the original town in the area, Linwood, was drowned under the reservoir. Now tourism is the main employer; all of the businesses in town revolve around the dam, reservoir, or river.

When I'd gone to talk to Steve Habovstak, the manager of Trout Creek Flies, the day before I started my run below the Flaming Gorge Dam, the river was just starting to come up, and Steve predicted that the high flows would crush his business. "High water really devastates us," he said. "We talk to the bureau about the economic impacts, but we lose $10,000 to $30,000 in revenue from high water, because people don't book, or they cancel." He said the fish are smart, they get weary from fighting the current when flows are up and down, so they'll hide in slack water and won't bite on flies. When the river is high and flushing, it's hard for guides to match the bug hatch that's going on downstream, too, so the accuracy of their flies—a big part of what customers pay for—isn't as good. The fly-fishing in the tail water below the dam is some of the best in the country, because the water is pristine and its temperature is regulated by the dam, but the varying water levels make conditions difficult to predict, and the flood stages that sustain the native fish are the opposite of the flows the trout like. "Spikes are bad," Steve said. "I've seen hatches tank in the past couple of years because of levels of fluctuation. We have this feeling that the dam doesn't care about our economy, they just care about power. We only have a small period of time to make money, you take two weeks out of it and that's a big deal."

That sentiment keeps coming up tonight: that some nebulous government entity is just out to make money from power generation and doesn't care about the people who really depend on the river. That frustration misses the subtlety of the fish program and the dedication of people like Tildon or Tom, who are trying to keep the river healthy, but it's still real. "The overwhelming attitude I've felt from the federal government is, 'you're a nuisance but I'll put up with this,'" says Ryan Weiss, who grew up in Dutch John and guides for Trout Creek Flies with Steve. "My business is suffering from this. I don't want to lose the human element. We can deal with Mother Nature, it's the apathy from the government, whether it's real or not, it's perceived, that's hard to deal with."

Finally, Heather Patno gets up to speak. Locally, she's often the target for the government hatred because she's the Bureau of Reclamation's point person. She explains the parameters the bureau is trying to stay within, as well as the historical and logistical reasons for its plans. She's working with hard targets in a flexible, unpredictable world. It's not just a question of enough water, it's how much and when, and how to control for the externalities of dams and the ways people choose to live in the desert. She is gracious in the face of the vitriol and tries to underline the connections. And slowly, as people talk through their grievances, and feel heard, the temperature in the room comes down.

When I leave, Ashley is talking to the Sheffers, and the camera is off. Aldis is getting advice from some of the ranchers about graphing predictions, the kinds of weather alerts they find useful, and how to send out more helpful forecasts and warnings. People's anger seems to simmer down after they feel like they're being heard, and when they see those on the other side as people. The hawk guy has stopped yelling, and Tom looks less nervous.

"I think it's going to get there, I just don't know how long it's going to take, or how many people's lives it's going to impact in the meantime," Kalynn Sheffer says, about the ways the fish program is trying to figure out and control flows. Heather says they've found that the best way to build consensus is to get people in the same room so they can hear one another's priorities, even if the conversations are shouted and painful at first. In-person interactions like these, where Tom can apologize for the destruction and explain why he's doing what he's doing, help to break down long-held stereotypes and deep-seated anger, but it's hard to see how that can happen on a broader scale. Even tonight, the group of people in the room isn't representative of the many more who depend on this stretch of river. It's heartening on a small scale and incredibly frustrating, writ large.

It's well past dark when I leave. I'm exhausted just from listening, and it's raining for the first time in weeks. The heat has finally broken. As I drive out of town toward my campsite, drops crystallize on my dust-speckled windshield, a reminder that it's not just how much water you get that's important, but when.

WHAT'S THE POINT OF A WILD RIVER?

The last major undammed tributary of the whole Colorado River system, the Yampa River, flows into the Green in Dinosaur National Monument, just north of Vernal. It drops out of northwestern Colorado, where it picks up water from Pat O'Toole's ranch, then runs through downtown Steamboat Springs, where tourists tube and fish. It meets up with the Green at Echo Park, Colorado, in the deep, golden canyons of the monument. Jack Schmidt says you can think of the Green as really starting there, at the confluence, because the inflow of the two rivers is almost equal. They become an entirely different river when they come together. The Yampa's warm, sporadic, sediment-heavy flow drastically changes the composition of the Green when it hits. Downstream of the confluence the river seems more untouched.

Part of the endangered fish recovery program's mandate is to use the Flaming Gorge Dam to make the Green look as much like the Yampa as possible, even just below the dam, because the fish depend on wild seasonal peaks. That attempt to return to natural flows, to mimic an undammed river, creates some of the drama that comes up in the Flaming Gorge Working Group meeting, because in the balance of uses, the fish carry more weight that agriculture or other industries. Protecting the Yampa has also become a political battlefield in the Upper Basin because it holds some of the last unallocated water, which plenty of people would love to divert. Once a river is dammed up or piped out, it's nearly impossible to bring back its untouched state, which makes the Yampa invaluable for providing a rare glimpse of an unaltered ecosystem.

The Yampa sparked this whole journey for me. The previous spring, I'd lucked into a last-minute trip through Yampa Canyon from Deer-lodge Park, in Colorado, to Split Mountain, Utah, just outside Vernal. It was organized by Friends of the Yampa, a Colorado-based nonprofit that has fought to keep the Yampa free-flowing since the early 1980s, when the river was threatened for the first time by plans for transbasin diversions. Part of Friends of the Yampa's mission is to bring dispa-rate water users together, and they've decided over the years that the best place to do that is on the river itself. At the put-in, water district managers, ranchers, and lawyers awkwardly shook hands and joked about how much beer they'd need to get through the trip. On the river, I listened to Jim Lochhead, who runs Denver Water, break down how he's trying to plan for the city's future through both conservation and storage, and Anne Castle, former Deputy Secretary of the Interior, talk about building treaties with Mexico. It was a group of people who were often at odds with one another, but they'd agreed to come along, and to talk, because they think the future of water needs subtle solutions, and that the Yampa is a vulnerable point in that future.

The trip taught me about the tangled legal jumble of water plan-ning, but also reminded me how much I loved being on the river. I'd been out of the guiding routine for almost a decade, pulled out of sea-sonal work by a full-time job, and pushed away, on an internal level, by an increasing fear of hurting someone, or myself, on the water. By the time I stopped guiding, I was nervous so often that the anxiety began to outweigh the appeal of running rapids. I loved water's graceful, com-pounding power, but it scared me, too. I'd started to obsessively notice everywhere a boat could flip or pin. My heart pounded and my voice went screechy any time the water got big, and I no longer loved the adrenaline I used to thrive on. My fantasies used to be different. As a rookie raft guide on the Penobscot River in Maine, I would stand on the edge of the Ripogenus Dam so I could watch the river run out of the turbines and disappear into a tight granite canyon. I'd trace boat lines through rapids with names like Exterminator in my head, trying to picture where I could punch through wave trains and where recircu-lating holes could flip a raft. I'd head for the biggest hits, until I flipped and swam enough times that I learned what it felt like to be held underwater to the point I wasn't sure I'd ever surface. I'd had clients go for the same kinds of swims, and I dreaded the raspy, big-eyed looks they always had when I pulled them back into the boat, choking on air.

The Yampa's unfettered runoff added another element of risk—you could never be sure what it was going to look like. The river cuts through shimmering, high-walled canyons filled with caves and strewn with sandy beaches. It's raftable only when the runoff peaks in the early spring, which results in stiff competition for permits. Its variability felt different from that of almost any river I'd ever been on. Once we were deep enough in the ditch to be cut off from any outside communication, we had no idea how much water was coming downstream. We'd watch the canyon walls for the telltale wetness of a bathtub ring in the morning, wondering if the river level was going up or going down. It was always changing.

That peak and spike of flows demonstrates the Yampa's wildness. There are a range of reasons why it hasn't been dammed or drastically changed. It's relatively remote: there's almost no urban growth along its course. The energy industry has minimal rights to water in the area. Agricultural users along the river's course have been staunch in their traditional irrigation practices; they haven't forfeited their water rights to cities, so not a lot of water has been transferred away from agriculture. But, perhaps most importantly, the endangered fish program, backed by the Endangered Species Act, leans on its flows and the habitat those flows help to create. Environmental groups, like Friends of the Yampa and American Rivers, have worked with the government to protect the Yampa and to make sure that its unregulated flows align with state and federal compacts, so it's not easy for new users to bite off a chunk of water. But because it's some of the last unregulated water in the West, there are often eyes on it, especially in dry years, and it doesn't have any concrete protection from future development. It's been threatened by proposed dam projects, oil shale and other energy development, and transbasin diversions. Plans for large-scale, expensive diversions that would pump Yampa water to cities on the eastern side of the Continental Divide surface every few years. The latest is the

Yampa River Pumpback,[1] a scheme designed to take 2,000 cfs out of the river and pump it to Colorado's Front Range, at an estimated cost of $3 billion. So far, these diversions have all faltered because of cost or regulatory blocks, but there's worry that the price could be right some- day, if water on the Front Range becomes expensive enough, or the city's supply dries up. If the rights to the water were sold off or its uses changed, the character of the river would be totally different.

The government has a designation for wild rivers. In 1968, as big dams were going up across the West, Lyndon Johnson signed the Wild and Scenic Rivers Act into law. It gave Congress or the Secretary of the Interior the power to designate rivers as "Wild," which meant that they were dam free, inaccessible by road, and unpolluted, or "Scenic," basi- cally the same, but accessible by road.[2] Once they've been designated, rivers are essentially untouchable. But because the act is so protective, its scope is small. It applies to less than a quarter of one percent of the rivers in the country. The Yampa is under study for designation, and has been since the 1970s, but it still isn't wild with a capital W. The designation process is long and fraught, especially when industries or other users have a say in what happens to the water, which is true of the Yampa. Its future always feels uncertain, and it's become a symbol of what we might stand to lose if every ecosystem is treated like a re- source to be fully consumed.

Because it's currently free of dams and major diversions, the Yampa is full of fish. It's prime habitat for the four endangered species, and it's also a stronghold for other, non-endangered native fish, like other species of suckers and chub. Tildon says that's partly because of the geomorphology. Variable flows build up braided, complex channels, which deposit the kind of cobble substrate the fish like on the bottom of the river, forming spawning bars and slack-water pools. Sediment washes in during flow stages, then erodes away over the course of the summer, bringing nutrients.

On my Yampa trip, we stopped on the cobbled bar that holds one of the last two Colorado pikeminnow spawning habitats in the Colo- rado River Basin. It came up quickly around a light-strewn bend, half grown over with grass and willows. The river flowed through the substrate, ankle deep. Tamara Naumann, the park botanist for the monument, was with us, and she explained why the bar was rare and important. The fish have habitat because the river hasn't been dammed, and the river hasn't been dammed in part because of the fish—because the recovery program protects their habitat. The Yampa's wildness intertwines with nativity.

Tamara has been working for the monument for decades. She's so embedded in the canyon that she seems like she's part of it: craggy and kind; practical but quick to laugh. Even her lifejacket is Park Service gray and green. It's her job to knit together the divided threads of ecosystem management and the current human-altered climate. She can't just manage for wildness by leaving the river untouched. To keep fish, insects, plants, and soil in balance, she's been a big proponent of initiatives like Anya's bug study. She's also responsible for a large-scale experimental tamarisk removal program. Invasive tamarisk, a shrub native to Asia that ranchers brought to the U.S. to stabilize ditch banks, has overwhelmed a lot of rivers in the Southwest. It spreads easily, and it's hard to get rid of because you have to completely destroy its roots, which snarl deep in the sandy soil. One of the only things that kills it is the tamarisk beetle, which eats the foliage and weakens the plant, but which is also invasive. After twenty years of study, and of weighing the risks and benefits of altering the landscape further, Tamara released the beetle into the monument to fight back the tamarisk. She says it was the scariest day of her life, because the beetle might have spread out of control, but instead it's cut tamarisk growth by half.[3] Her strategy is similar to what the fish recovery program is doing: they're trying to bring back the native fish in a human-managed but nature-reflective way. People, climate change, non-native plants, air pollution, and more have distorted the river. Leaving it in its current state would mean abandoning it to inadvertent human impacts. The peak flow coming out of Flaming Gorge is an intervention, as is keeping the tamarisk at bay, and making sure nothing happens to the spawning bar. On the river, someone says they think protecting the undammed Yampa feels important to so many people because almost nothing is truly untouched anymore.

Because the Yampa has the capacity to act as a naturally functioning river, endangered fish can still find a niche in it. And Tildon says that because of their federal protection, a certain amount of water has to keep coming downstream. "Boating isn't going to protect the river, there's no law that says you have to protect rafting. I think fish have the most likelihood of keeping something out," he says.

On the Yampa, our group floated in a loose gaggle of yellow boats, sliding through the sunbaked river miles slowly. We stopped to scout before big rapids like Warm Springs, which was formed by a massive landslide in 1965. We climbed up into caves that once held hideouts for outlaws, and we hiked up to overlooks that gave us the scope of the striped layers of the golden and green canyon. At night, around the campfire, we passed whiskey, made up songs about the trip, and talked about the future of the river. We tried to explain to one another why its wildness felt important to us, what was different there, in that particular canyon, from other rivers we'd been on, and why all of us had shaped our lives, to some degree, around water.

Eric Kuhn tells us the river's tangled history. He's the general manager of the Colorado River Water Conservation District, and he's known as an expert and a sage across the Colorado River Basin because he has a rare, unbiased ability to hold all the conflicting uses in his head. He's also good at explaining the pressures on the river and how the structural deficit will play out over the long term. He told us that the Colorado River Compact and the way states' water rights define beneficial use make it harder to keep a river wild than to tame it. In all the basin states, flows for recreation are contentious and not easy to come by, and ones for species conservation or environmental protection are tricky, too. That's where groups like Friends of the Yampa come in. Because there's no clear beneficial use in not using a river, they have to lobby to keep a river untouched. To underscore why the river is so important in its current state, they're trying to tie it to ecological benefits and the economic uptick that recreational uses bring to communities. That's why they brought us all to the canyon.

Currently, the value of water is tied up in the direct profitability of its controlled flows. Ecosystem values are harder to quantify. Kent

Vertrees, our head guide on the Yampa, who works for Friends of the Yampa, sits on the Colorado Basin Roundtable, a group established to bring regional perspective to the state's water plan, which outlines how water will be used in the future. He says they initially had to fight to explain what a wild river brought to the area, and to the wider river basin, especially in the agricultural Yampa River valley. "People said it's a waste of river."

People don't like the idea of setting aside water for fish or paddling because it renders it unusable for agricultural or municipal uses, and beneficial use is basically holy. "We're trying to convince the state of Colorado to push all this water downstream, which means it doesn't get used in the state," he says, "I would love to see it, because it keeps the Yampa wild and it would provide water to the compact obligations, but people don't want to have a sacrificial lamb. But do we need to de-water every system? When is enough enough? Why not have one river left that still has that capacity to be wild."

It's a polarizing question in the moral universe of river management, and really in any kind of resource use: What is the value in not using something, and how can you quantify that value when it's less tangible than that of farming or municipal uses? How does society settle on a hierarchy of uses? Kent doesn't see allocating every single drop as a smart way to manage long-range needs. "Who knows what's going to happen in the future?" he says. "It could be a major wet cycle. But you have to think about the scariest cycle, especially if we manage water like we do today, if we keep spraying treated water on our lawns, and flood irrigating."

Tildon calls the Yampa the lifeline of the Green. He says it keeps the river alive, but it's also the heart of the hard-to-answer questions that everyone who works on the river is trying to answer. Do we need pristine wilderness to sustain wildness? Is that even a reasonable goal? Is having a sliver of untouched river, even if it's just a fractional amount of water that ends up evaporating off later in the system, worth protecting for the sake of what it connects?

When he tries to show that leaving the river wild is intrinsically valuable, Kent has to wrestle with those questions, too. So does Tamara, when she tries to keep the ecosystem within a range of what it might have been without human impact. The Yampa gives us a slice of what things might have been like if our predecessors hadn't changed the river, on purpose or inadvertently. They say we can learn from it, that it gives the natural world some space.

When we entered the Yampa's confluence with the Green, we came in silently, holding our breath. I rowed that section, trying to dip the oars as quietly as possible, not wanting to interrupt the moment or push us downstream too fast. The river bottom at the confluence is called Echo Park because the towering sandstone walls bounce back sound, and it is unfracturably beautiful. When we hit the bend where the cooler, clearer Green comes running in, we broke our silence, yelling, our voices echoing off the walls.

ONE BIG FISH TANK

The road south of Vernal that runs through the Ouray National Wild-life Refuge, to the fish hatchery within its boundaries, is hot and dry. It parallels the river, and as soon as I'm away from the banks, the land-scape is dusty and barren. It's amazing to me how dry it can be right next to the water. It's arid and vaporless, with the heat rising up and sucking energy out of the air. When I pull up to the hatchery, it, too, feels empty. It's dark inside, and tanks full of fish are bubbling, but it takes me a while to find another person.

I wander until I hear voices at the end of a hall. Matt Fry, the fish bi-ologist I'd planned to meet, is having a conversation with his coworker, Trenton Thompson, at the urinal. I hover, awkwardly, until they come out, and we walk into the dank warehouse that holds the tanks.

Matt says it's not an ideal place for a hatchery. It's often this hot, and the water they pump in is alkaline and hard on the fish, but we're in Utah, where the conservative state legislature isn't prone to allocating resources to federal programs. In 1996, when the Ouray National Fish Hatchery was started to augment the fish recovery program, the land within the wildlife refuge had already been set aside for Fish and Wild-life use, so they didn't have to fight to house the unpopular program here. He says they do the best they can to keep 30,000 fish alive here in the winter, even though the mineral-heavy water necessitates mul-tiple filters, and outside, the predators, including birds and salaman-ders, are fierce, circling the thirty-six production ponds full of adult suckers and bonytail.

Tom Chart defends the policy that protects the fish, Tildon Jones carefully monitors their life cycles, Tamara Naumann fights nature to maintain their habitat, and here, Matt, who has dark hair, dark teeth, and the ubiquitous khaki government uniform, is building back the decimated fish population. Even if the habitat is pristine and there are protective legislative measures in place, none of that matters if you don't have any fish.

The hatchery program's initial goal was to rebuild the razorback sucker population. In the 1970s and 1980s, after Flaming Gorge Dam went in, fish biologists realized they were only seeing aging adult fish in the river—they couldn't find any juveniles. Razorbacks can live for forty years, but because of the dam there were no flooded bottomlands for spawning habitat, so the fish weren't making babies. The river had become too narrow and deep. In the late 1980s, Fish and Wildlife started an experimental artificial habitat in two ponds within the refuge. They captured native fish and started breeding them, keeping the fry in tanks and building up a hatchery population. Some of the brood stock is still from that original batch, hatched in the summer of 1989, and Matt points out the pond they live in as we walk past. Despite the oppressively hot day, the blue-lit building is damp and cool inside. "I like to think of it as one big fish tank," Matt says, as we wander down the rows.

When the hatchery started in earnest, in 1996, it was tasked with breeding fifteen thousand twelve-inch razorback suckers each year, which Fish and Wildlife would then release with PIT tags so they could be tracked. At first, the results were dismal: hardly any of the fish lived after they were released into the river. But the biologists kept experimenting, tweaking any portion of the life cycle they could to try to help the fish survive. Some of those little changes made a difference. Matt says the twelve-inch fish had only a 30 percent survival rate in the wild, but the biologists found that if they let them grow to fourteen inches before they were released, the survival rate went up to 70 percent.

In 2010, they started to breed bonytail, which were essentially extinct in the wild. Now they raise six thousand razorbacks and ten thousand bonytail each year. They shifted because the razorbacks they released were doing relatively well in the river, while only about 2 percent of the bonytail were surviving in the wild, so they needed more help. Those hatchery fish have to live long enough to spawn so they can build their ranks back up.

Repopulating the river is not a simple matter of breeding more fish and letting them go, especially because there were so few in the river when the effort began. One of the tanks holds twelve wild humpback chub from Desolation Canyon, just south of us, that have been living monitored in the hatchery since 2009. Matt says that chub don't do well in captivity. Initially, the fish were so stressed that they were constantly trying to jump out of the tank, until the biologists gave them a tarp to hide under. Matt pulls it back, to show me, and the thick, dark fish start to thrash frantically.

The hatchery has kept the humpback chub to see how they do in tanks over the long term and to better understand the fish's life cycle, even though it hasn't been able to breed them. It's part of a series of adaptive experiments the hatchery is doing to figure out how the fish live best in captivity and how they can hold onto as many of their wild traits as possible. Matt says he's trying to manage for a baseline he doesn't fully understand.

Hatcheries and the fish they produce are contentious. The traits that make hatchery fish successful, like aggressiveness, are not necessarily the same qualities that make wild fish thrive. Fish bred in hatcheries tend to be big and mean, combative and genetically dilute, and they don't have the flight reflexes and learned shyness that helps wild fish survive in their native habitat. It's a tricky question: are some fish, even if they're marginal, better than no fish at all?

Matt has obsessively tried to reduce the hatchery influence and to keep the fish as close to wild as possible. He's constantly experimenting with ways to tweak the system to make the hatchery-bred fish more resilient. But I can see the difference between the skittish, wild humpback chub and the hatchery-raised bonytail, which school up for feedings.

A snake slithers off into the grass as we walk up from the warehouse to look at the outdoor ponds, which sit gridded on the adjacent hillside. In addition to keeping the fish healthy and growing, the biologists have to fight off the predators, because the fish have nowhere to hide. In 2007, almost all of the fish in the outdoor ponds were picked off by herons, which swooped down on the young of the year. "It was like Keystone Cops trying to keep them off," he says.

At first, the concrete ponds look dark and dead, but when we get to the edge, I stand on the diving board–like feeding platform, look down, and see thousands of two-inch-long bonytail twizzling beneath my feet. Matt says he's experimenting with double cropping in this pond—putting both fish species in the same pond. They seem to do fine; the razorbacks feed on the bottom and the bonytail on the top. "It's a science," Matt says, "but it takes some finesse too. Their livers are fatty from the fish food. You look at their fins and gills, see if they're rubbed off. They get costia and other parasites."

Matt has worked at the hatchery for a decade, and he's committed to it in an unrelenting way. He has an alarm that tells him when there's a problem with water circulation, or if the heaters go off when he's at home at night. He's gone rushing back to the hatchery at midnight in midwinter snowstorms to defrost pipes and keep sensitive fry at the right temperature.

Despite his fanaticism, Matt didn't set out to be a fish biologist. He was in college at Northern Arizona University, gearing up to study elk in Montana, when he had to take a fish biology class to fulfill a requirement. He was assigned two fish species to profile for a paper, and one of them was the bonytail. He couldn't find any information about the fish, except for a mention of the rotenone poisoning below the Flaming Gorge Dam in 1962, which killed off the native fish to make way for sportsman-friendly rainbow trout. "How could we have done this?" he remembers thinking. He says he's been obsessed with fish ever since.

Matt's narrative is mainly boilerplate as we walk through the hatchery: he runs me through the history and the math, telling me little stories when I ask. But when I ask him why he thinks it's important to protect fish, and why he spends his days, and sometimes his nights, blowing heaters at frozen pipes and screaming at birds to protect measly fish fry, he cracks a little. He takes me back into the break room, where leftovers sit in the fridge next to sacks of fish food and everything smells ripe and overfermented.

He says he's constantly worried about the future of the river, and about what might happen to the slice he sees—the fragile endangered fish—if there are water-supply shortages. He thinks the Colorado River Compact could be broken if things get dry enough that there isn't enough water to fulfill it, which would mean that Upper Basin flows would be reduced and the fish would suffer. "When push comes to shove, and it's people versus fish, people win," he says, running his hands over a set of vials on the counter. "To me that's a nightmare. It's a paradigm shift right now, and it's a painful time." He says he's worried that the hatchery's water or funding might be cut off, and that the fish populations would disappear without the steady attention the biologists are giving them. Fear, especially fear of the compact being broken, or the Lower Basin making a call on the river and demanding its allotted water—the 7.5 million acre-feet a year that the Upper Basin is committed to provide—seems like a constant undercurrent in any way the water is used, even though those things have never happened before. So many people I've talked to feel like their livelihoods, or the things they hold important, could easily slip away.

The future of the fish feels tenuous here, even with the constant attention. Matt says Fish and Wildlife is trying not to lose biodiversity and to hold the underlying threads of the ecosystem together, but it takes a huge amount of work. The fish recovery program has population targets, and those populations have to be self-sustaining. The fish have to complete their life cycles in the wild, and that's the real challenge: to translate what's happening here, in highly controlled tanks and ponds, to the volatility of the river. Recovery isn't helpful if it happens in isolation.

The biologists have to plan for the threats that brought the fish here, such as human impacts and changing climate, so that they can sustain themselves into the future. Their management takes a multitude of forms. Even though it feels like Matt is on an island out here in an easy-to-miss pocket of the desert, obsessively recording fish skin fungi, he wouldn't have a baseline population of fish if not for Tom's work, or Tildon's. They wouldn't have pikeminnows if not for the Yampa spawning bar. Keeping the fish alive feels like a game of Pick Up Sticks: the bottom could be pulled out by any number of things—drought, or defunding, or spills from upstream drilling.

When I watch Matt walking the aisles between the fish tanks, glancing in at the teeming tiny fish, knowing that most of them won't survive, it feels futile, but he's cleaved to the work. I'm not sure what he would do otherwise, so until it collapses, he'll keep trying.

RECREATION

9,180 CFS

THROUGH THE GATES

The neck-craning canyons of the Gates of Lodore spring out of no-where, at odds with the rolling buff of Browns Park. I should be used to geologic shifts by now, but I'm still smacked every time the land-scape changes. After the mellow undulation of the park, the walls of the canyon pinch back in, and the red rocks of the Uinta Mountain Group fold up, pushing skyward. The water picks up speed as soon as we get on the river, and we paddle into a stiff, persistent headwind. You're ruled by wind and light in the canyons, and the sun diffuses as we get deeper into this one. I keep spinning my face skyward, trying to get a look at the top.

The evening before, I'd showed up at the O.A.R.S. rafting outfitter's boathouse in Vernal to prep for a commercial river trip through Lodore. River permits for the forty-four-mile-long section are among the most sought after in the country because those sky-high canyons hold sus-tained class IV rapids. I'd talked my way into tagging along on the trip so I could see how the rafting industry plays into the mesh of water use.

We sit around picnic tables in the waning light and introduce our-selves, explain why we came here, and describe what kind of time we've spent on rivers before. The rest of the trip comprises two families—brothers, their wives, and two kids each—and a group of retirement-aged friends from Grand Junction, Colorado. I hadn't an-ticipated being the oddball on this section of this trip. I'm unsettled by the sudden shock of social structure, and by having to explain myself. I'm not attached to any group, not one of the guides.

Our trip leader, Russell Schubert, gives us a rundown of the plan for the next four days. We sign waivers, and he sends us off into the night with the promise of coffee by 7:00 a.m. I head back to my room at the cheapest hotel in town, suddenly lonely after being alone for so long. I've been on the river for more than a month now and I've hit a weird halfway point of melancholy, unattached and trying to figure out what I've learned. Part of what I like about being on a river is that it's specific to a time, place, and group of people, but right now I don't feel like I have any of that. I've dropped into these people's vacation, and I'm in the middle of a trip I'm struggling to explain. I feel unmoored and out of place on this river I'd started to claim as my own, and on the edges of the guiding world that used to be so familiar.

In the morning I slip on my Chacos and board shorts, make a soggy waffle in the Super 8 lobby, and head back to the boathouse. I've grown accustomed to quickly packing my own boat and bags, so the Tetris of tourist gear feels unwieldy and slow. The Green comes into Vernal from the northwest. In four days we'll take out twenty minutes from town, but the river bisects Dinosaur National Monument and there aren't any roads across the monument's steep canyons, so we have to circle a few hours around to the put-in. We drive into Colorado, across the top of the Tavaputs Plateau, coming into Browns Park, where I last took off the river, from the east instead of the west.

We stop in Dinosaur, Colorado, just over the state line, to load up on booze. It's not even 9:00 a.m., but the liquor store, one of the few businesses in town, is open and full of the grease of river trips: boxed wine and canned beer. The store knows its audience, and knows that most paddlers are coming from Utah, where beer is weaker, because there's only one road through town. Route 40, which runs from Vernal through Jensen and across the state line into Dinosaur, is a stretch of trailer towns and vacant lots. It's always been empty country, but the oil downturn, which is so present in this part of the world, makes it feel ghosted and abandoned, like it was chainlinked and left for dead.

There is a crush of rubber and gear at the boat ramp—dry bags strewn outside the pit toilets, sun block everywhere, as everyone scrambles to adjust their layers and rearrange their kit. The guides are at the ramp already, strapping down coolers and adjusting oars. Launch day is always messy, even on a commercial trip where the guides have done the hard work of rigging, because no one quite has their rhythm yet. We all double-check our snacks and water bottles, tighten our sandal straps, and try to use the bathroom one more time. The group that launches ahead of us forgets a watermelon on the concrete ramp, and so we pick it up and take it with us, following them downstream.

Even though I feel awkward about being the odd one out, I'm antsy to get back on the river, especially after a stretch of hot, dry days in Vernal. I love the hustle and flow of a river trip, the pattern matching of linking your days to the flush of the river, the forced efficiency of traveling by boat. Time is compressed in the canyon because you can only go as fast as the water. You succumb to the speed of the river once you push off from shore, and there's a rhythm to the days: the structure of the setup and teardown at each beach, the careful packing of rafts.

The river moves backward through time here, downcutting into progressively younger rock. Lodore's canyon is deep and dark red at first. It's vivid with the waxy green of box elder along the benches and lighter, wispier tamarisk lining the shore. Powell called the canyon a dark portal when he first saw it, from his camp the evening before he and his men paddled through.

Flaming Gorge is still releasing full flows to keep the reservoir from overtopping the dam, so the water is high, and we move quickly into the canyon despite the fierce headwind. The first major rapid, Winnie's, is flushed out and riffly from the push of water, barely even there. Rivers change when the flow varies, which can be challenging for the guides, especially in sections like these that they might see only a few times a year. When we stop to scout Disaster Falls, my guide for the day, Bob Brennan, tells me he's never been down the river at this level before. From the upstream eddy where we pull our boats over, I can see a mushroom wave cascading over a frothing pourover and hear the river thrumming through the canyon below. We hike a skinny, foot-beaten trail downstream to get a better look. Once we're beside it, the rapid is gnashing and violent, with a froth of diagonal waves pushing toward compressive holes, spitting spray. The guides are on edge. Kerry Jones and Bob point and murmur to each other, tracing the tongues of waves with their fingers, predicting where the water will push their boats. Jamie Moulton, who is a Grand Canyon guide most of the year, is biting her nails, staring hard.

I'm not naturally patient, but I can watch water move for a long time. Fluid dynamics are the same in any river—rocks make eddies, holes, and waves, depending on flow and constriction—but every river, at every different water level, pulses and reacts in different ways. Standing on the shore at Disaster, watching the river pillow up and crash through wave trains, I can feel a familiar tightening in my chest.

I know athletes—skiers and runners, mainly—who visualize their events, picturing themselves making every turn, to prepare themselves. Scouting whitewater is that same kind of meditation and projection, but it's more visceral, because you can see the muscle of the river, and all the ways it can mess you up. From the bank it's just as easy to visualize what can go wrong as what can go right.

River running done right is all grace and ease, a few clean oar strokes here or there, minimal time spent fighting the current. But in big water there are a million ways to miss your mark and find yourself pulling against the waves, fighting hard. We'd passed an Outward Bound group of scraggly teenagers and tough-looking guides walking back upstream while we were walking down to scout, and they start to run through the rapid while we're standing on the banks. Their trip leader rides the tongue of the current to the edge of a hole and then pulls back away from it gracefully and powerfully, skirting the edge without any wasted motion.

Powell lost a boat at Disaster Falls, and so did fur trapper William Ashley, who came through the canyon in 1825, looking for beaver. Much of the history of recreational river running cascaded out from their explorations on this stretch. In the 1930s, before the dam went in and before inflatable boats like the rafts we're in now were invented, the early wild men of paddling took wooden boats through the rapids of the Green—first to prove they could run it, then to explore it, then to make money by bringing clients into the canyons.[1] In the early 1930s, Vernal-born carpenter Bus Hatch, who was known for his drinking habits and disregard for his own physical safety, started running the

Green through the Gates of Lodore in wooden boats with names like *What's Next* and *Who Cares*. He modeled them on Galloway-style boats, the first designed to be rowed backward, which were named after their designer, Nathaniel Galloway. Bus learned about Galloway boats after he helped Nathaniel's son Parley jump bail in the Uinta County Jail. But Bus was more of an opportunist than a criminal, and after gaining some recognition for taking friends boating, he realized that strangers would pay him to bring them down the river.

Recreational river running grew in parallel with both the explosion of dam building across the West and a mounting interest in environmental protection. The industry changed after World War II, when boaters realized they could recycle military pontoon boats. The new rubber rafts were cheaper and more resilient then the wooden Galloway-style boats and could hold more people and gear. In 1953, Bus Hatch was given the first National Park Service river concession permit in the country and allowed to run commercial trips through Dinosaur National Monument.[2] By the mid-1950s he was taking hundreds of people through the canyons of the Green and exploring rivers in Idaho and Arizona. Other companies sprang up in Vernal after that, and Bus's son Don eventually took over the Hatch operations in Utah.

The river still feels wild and lonely now, but there are guidebooks, tourists' GoPro videos on YouTube, and a sense that we're retracing other people's steps. I envy the way those early paddlers explored these canyons. The history, and the myths about people like Bus, make me nostalgic for a time I wasn't part of. I feel like I missed out on the best, most untamed days of river running.

I'm jealous in part because I know how wild I felt when I first started boating. In May, my first summer on the river, my mother dropped me in a rafting company's parking lot with a PFD, backpack, and tent, and told me to call at the end of the season. I didn't have a car or a cell phone. I was eighteen and had been in a raft once. I spent the next week in training, running the icy twelve-mile stretch of rapids below the Kennebec River's Harris Station Dam as many times as I could, learning how to flip boats, how to find downstream vees, and how to draw and pry rafts into line. I slept with my soggy base layers in my sleeping bag, hoping they would dry out overnight. In the morning, the other guide trainees and I would run our frozen wet suits under a hose to make them pliable enough to pull on. I learned that rapids smell like the rock they run through: cool granite or minerally sandstone. I loved the calculated risk of it, the subtleties of slipping through the bubble line at a top of an eddy, and the gut drop of tipping over the top of a wave train. I became obsessed with pushing myself in bigger, more complicated rapids because it made me feel like I had power. Once, on my day off, I jumped into the river at the dam with a boogie board to see what the river would do to my body, instead of a boat.

When I moved to Colorado after college, in my tamped-down, modern version of manifest destiny, I spent summers pushing boats on both sides of the Continental Divide. I worked on the Eagle and the upper Colorado, which ferry water toward the Pacific, and the Arkansas, which flows east out of the Rockies toward the plains. I'd started rafting because I had an antsy, unfettered desire to be outside, but I got sucked in because of the charge and the community—the slow rolling way that being on the river binds you to people. After I fell in love with paddling I started to learn about the multi-threaded controversies of water use and all the nonobvious ways that rivers are important. Guiding tourists was my line in. For me, the world of commercial river running, even though it sometimes feels contrived, is an important part of river access.

Here on Lodore, that sense of value feels clear as we slip into the day-to-day routine. Coffee and breakfast, reloading the boats, the first flash of water to the face. Lunch, and an eventual ache from paddling, then beers and books on the beach at night. Everyone starts to open up because they have time and space. Teenager Thomas Griffith sits next to me by the fire and tells me he might want to be a writer someday. Sometimes we just sit around sun-dazed, not talking, sipping Tecates and eating cheese.

An O.A.R.S. trip isn't cheap. Being on the river like this—and, really, being on the river in general—is an expensive, hard-to-attain privilege. If you don't come on a high-end guided trip, permits are tough to come by, gear is expensive, and you need to have some level of skill. But I care about water because I cared about running rivers first. I wonder, sitting around the circle with Thomas and his sister Nora, or watching the littlest girls, Josie and Edie Hughes, squeal in rapids, what they'll remember from this, and what will carry weight for them.

The morning we paddle into Echo Park, where the Green joins up with the Yampa, the upstream breeze is so stiff that when we stop against the walls of Steamboat Rock to see if we can make it echo, our cries of "Spaghetti! Watermelon! Hellllllooo!" get pulled off by the wind. We're spit out the mouth of a narrow canyon, with Steamboat Rock, a slab of the Weber Sandstone that looks like the turreted towers of a massive ship, monolithic on our right and the Yampa flowing in from the left. Echo Park is hydrologically pivotal; the river changes here, and when we float around the corner, the Yampa is pumping, swelling the Green to twice its size. Even though it only flows in a short spring snowmelt pulse, the Yampa brings in about the same amount of water that the dam-controlled Green does all year. The confluence is also breathgrabbingly beautiful. The Green hairpins in, greener and clearer, picking up the warmer, muddier Yampa before it makes the bend back around Steamboat Rock. The point between the Green and the Yampa is a wide alluvial fan, topped with a thick bench of pale sandstone. Last summer, when I'd floated in from the Yampa side, silent except for the slip of the oars, I'd been awed by the glimmering towers of the park.

This area was almost underwater once. It was slated to become a reservoir, filled up to the geologic folds of the Weber Sandstone and Morrison Formation by a dam two miles downstream. The Bureau of Reclamation proposed the Echo Park Dam in 1941. By building one dam in Whirlpool Canyon, just downstream of the confluence, and another slightly smaller one downstream of that, at Split Mountain, the bureau would have created a 6,400,000-acre-foot reservoir flooding the Yampa and Lodore canyons well upstream into Dinosaur National Monument, which had been designated in 1915. In the early 1950s, Congress planned to include that reservoir in the Colorado River Storage Project.

In 1951, the Sierra Club decided to fight against the two dams. Its members were still mad about the 1923 damming of the Hetch Hetchy Valley in California's Yosemite National Park and didn't want any more water storage in national parks. In 1952, the same year he became the club's first executive director, David Brower took a trip down the Yampa with Bus and Don Hatch. He said he'd never had an equal scenic experience.

The Bureau of Reclamation was pushing hard for the Echo Park Dam because it had calculated that a reservoir there would lose less water to evaporation than other proposed downstream reservoirs, but Brower didn't buy it. He did some scrap-paper calculations and determined that the bureau's evaporation estimates were wrong. In 1954, he testified against the dam in front of a U.S. House of Representatives subcommittee. He dragged in a chalkboard to show them that the bureau's math was faulty. Galvanized by his time on the Yampa, he became devoted to protecting rivers.[3] Brower had worked in book publishing before he turned to conservation, and he leveraged media to raise awareness about the fight against the dams. He started a letter-writing campaign, commissioned Wallace Stegner to edit a book called *This Is Dinosaur*, and paid the Hatches to take other prominent politicians down the river.

His campaign worked. In 1955, the Echo Park Dam was removed from the Colorado River Storage Project. "When my father and his faction won, it was the first time in American history that a group of citizens had stopped a big government project," Ken Brower, David's son, said in a documentary about his father's trip.[4] "The victory brought the Sierra Club and my father into national prominence and it served as a template for his successful fight, a decade later, against two dams proposed for Grand Canyon." Brower's galvanization of far-flung voters became a model for environmental organizations. Whitewater boating became an industry here on the Green, but it also shifted its focus to recreation and conservation instead of just exploration.

Because of Brower, the river-smooth sandstone towers of Echo Park are still visible, shimmering pink in the summer light above the suspended sediment of the river. We stop for lunch on the downstream side of the confluence with the Yampa. The wind lets up, and we make sandwiches on a sandy beach, standing ankle deep in the rushing water.

Places are complicated things to love. Conservation is often a job of making bargains, of trying to gauge what to give up and what is irreplaceable. Brower learned that you can't save everything. He said his great regret in saving Echo Park was that he put all his chips in this canyon and wasn't able to save Glen Canyon, too. Because Echo Park remained untouched, Glen Canyon Dam was built downstream, along with the Flaming Gorge Dam upstream, to shore up water storage for the growing West.

Brower believed in the value of beauty and awe, and so do we, even though we all struggle to articulate it when we sit around the fire at night telling stories, strangers now bound together by days in the canyon. Brower continued to protect wild places for the rest of his life, and I wouldn't be on this river if I hadn't learned to love a different one years ago.

WHAT IS IT WORTH?

Mornings in the canyon, the light is clear and the cliffs are more orange than red. In the neon, just-past-dawn glow, the river looks higher and browner. We only have four river miles to cover today—a short day on the river—so we hike up to an overlook before going downstream. Clambering up onto a rocky knoll to look back upriver at where we've come from, Gabriella, one of the women from Grand Junction, calls it a cathedral.

The canyon was formed by downcutting, the slow, persistent erosion of ancient rock. There are twenty-three rock layers within Dinosaur—it has one of the most complete geologic records in any national park. Fault lines slice through the landscape, and the striations in the uplifted rock dip and warp. There's a 500-million-year-old geologic unconformity between the Uinta Mountain Group and the Madison Limestone below it. It's enough to make you woozy when you think about the compression of time.

"When you're in water that big it's like a landscape," Noah Hughes, one of the dads on the trip, says once we're back on the water. He's in a kayak, and the rapids move by at eye level when you sit that low in your boat, but I can feel the way the river subsumes everything else in the raft, and off the river, too. We've hit the point in the trip, a couple of days in, when things get loose and the world outside the canyon starts to feel irrelevant. Chris Clark, one of the guides, has downgraded his wardrobe to a sparkly tiger T-shirt. My hair is a matted, silt-filled braid. We all pee in front of each other.

In the afternoon, we hike up to Rippling Brook Falls, climbing stairs carved into the ledges to a cup of a side canyon. From up high, I can see the curve of the river and the fins of red rock exposed where older, softer stone has eroded away. We take makeshift showers under the flow of the falls, screeching in the immediacy of the cold.

On the walk back, Kerry Jones, who has been a guide since the early '80s, hands me three juniper berries. He says it's part of his tradition after running Hell's Half Mile, which we'd paddled that morning. A lot of the guides are superstitious. Bob splashes his face before every rapid and thanks the river spirits after. My tie to any kind of spirituality is tenuous, but out here I can believe in the rituals and good luck talismans. In the concentrated world of a river trip, everything becomes a pattern: the placement of the groover, the call for coffee, the rhythm of water slapping the side of the boat as we push downstream. I don't want to take any chances and mess up the good things.

The wind picks up and the light on the cliffs turns the water pink as we make our way down the side canyon back to the beach. I dip in the river to cool off, and everything feels a little silty again. I want to hold onto this whole day: the color of the canyon, the ease of moving through it.

Days like this have driven conservation since Brower's time. Advocacy is a crucial part of keeping water flowing and keeping the river sustainable, but paddlers don't have a legal stake in water rights. Rafting companies like O.A.R.S. are dependent on consistent minimum flows for their livelihood, but they don't get a say in how much is released from the dam and when. A week like this one, with continuous high flows, is a helpful bonus for the outfitters.

Historically it's been hard, if not impossible, to secure water rights for recreation because in some states, including here in Utah, paddling is not considered a beneficial use of water. In the 1980s, Dinosaur National Monument filed for Yampa water rights in Colorado to guarantee instream flows for fish and boating. The request was rejected by the Colorado Water Court on the basis that the monument was designated specifically for its paleontological quarry full of dinosaur fossils, which didn't require water. The monument was barred from filing for similar rights in the future.[1]

There's some recent resistance to that, and a growing body of economic research that is quantifying the explicit value of recreation, healthy waterways, and instream flows for the sake of paddling. Groups like the river-focused business coalition Protect the Flows have formed to try to make sure the recreation economy—which the Outdoor Industry Association estimates generates $646 billion yearly in national consumer spending[2]—has water access into the future.

These groups are arguing that water-dependent industries shift and that the country's economy has moved away from small-scale agriculture, so traditional beneficial uses don't always need the most water. Underscoring the steady positive financial track of recreation can be a provocative argument. In Vernal, which has a long history as an energy industry town, there's palpable animosity toward boaters. If you're a guide, there are some bars you don't go into. Boaters have been pegged as hippies and environmentalists who are anti-development, even though they started the tourism industry in town.

Craig Mackey, the director of Protect the Flows, whom I met on the Yampa, says they've found that recreation within the Colorado River system generates $26 billion each year, outpacing farming revenues by 14.6 percent.[3] He and others say that's an important perspective for future water planning. Water managers need to look at how water affects livelihoods and lifestyles, including how and why people are moving to the West. We're past the days of homesteading. Almost no one is moving west to ranch or farm—"Ag in Arizona is 1 percent of the economy," he says—but lots of people are moving to Salt Lake City or Denver because they want easy access to recreation. The challenge is planning for both increased urban water use and recreational instream flows while still working within the system of prior appropriations. "People want to live in certain places because of the water," Craig says. "The biggest issue is that for 150 years we've done a really good job of taking water out of the river system, all for very good reasons, none of which are going away. We've done a very good job of harnessing the river and allowing the West to bloom, but in the twenty-first century we have economic reasons to have a river itself. The challenge is that the water system isn't set up to look at that world."

He wants to use that economic argument to back legislation that protects recreational water rights. Colorado, for instance, passed a law in 2001 that allowed water rights for instream flows specifically for paddling, called recreational in-channel diversions.[4] Those rights have to conform to highly scrutinized standards, and they don't apply across the basin, but they're a starting point to give paddlers a bit of agency over flows. Craig hopes his coalition's work can help faltering rural economies in places that can take advantage of the recreational potential of rivers, like Vernal. But any kind of change in that realm is slow and clunky. To make it happen, there needs to be a social switch in how water is valued, and in the norms around how it's perceived.

Here in the Uinta Basin, the drop in oil prices has crushed the local economy. It's incongruent with the rest of Utah, which is growing, and that makes the downturn hurt more. Locals tell me they're frustrated with the government, with regulations, with environmental groups, which they see as putting their jobs in jeopardy, and with outside liberals, especially the ones who breeze through for their high-dollar river trips.

In Vernal, if you're a liberal or a paddler passing through, you can expect to pay a buck extra for your drink at George Burnett's I Love Drilling Juice and Smoothie Cafe. George is Vernal's local celebrity. A former infomercial star, he now owns a seat cover shop that also houses the smoothie bar and a nail salon on the east end of Vernal's main strip. When he moved to town in 2008, he decided he needed a marketing hook for the store. Drilling was booming, and most of his customers worked in the oil fields, so he decided to stand on the corner holding a sign that said I [big red heart] Drilling. He quickly became the most recognizable person in town. I see him on the corner every time I go by, and want to talk to him, but I'm not sure how to begin. When I finally approach, he waves me in, saying he's never seen me before, and we start chatting. Cars honk as they cross through the light, and he waves his sign, grinning. He's mustached, handsome, and incredibly friendly, what the Marlboro man might look like if he existed on a diet of green juice and conservative values, which is what George tells me he's all about. I tell him about my trip. He loves rafting and being outside, he says. He calls himself a hippie, sort of, and he doesn't see any conflict between that and standing on a street corner gushing love for fracking wells. We're constantly interrupted. People, including an out-of-work oil-field worker, stop by to take pictures with him, and they earnestly thank him for his activism.

It makes me realize how much I've been making assumptions about the groups I consider the bad guys, even if I don't really know what their intentions are. I sneer at drilling trucks when they drive by, even while I'm trying to listen to both sides, even while I'm burning gas. I don't necessarily agree with George, but standing on the corner with him, and hearing him talk about how holding his sign feels like supporting the community, I feel hypocritical and torn as I nod in agreement.

So in the same slow, case-by-case ways that Tildon is explaining the eradication of non-native fish to anti-government fishermen, groups like Protect the Flows are trying to change that conversation on conservation. They're consciously focusing on the economy, instead of habitat or boating access, because it doesn't hit a political hot button in the same way that talking about the environment does. They want to make tourism seem like a value, not a sore spot.

These torn-up deserts dotted with drilling rigs could benefit from a stable, sustainable economy. Vernal has been seesawed up and down by the volatility of oil and gas prices. People like Craig are trying to spread the word that recreation can provide long-term jobs, but it's hard to change long-standing frameworks.

On my second-to-last afternoon in Lodore, I ride in Garth Butcher's boat with some of the people from Grand Junction: Gabriella, Nikki, and Kyle. After the bleached openness of Echo Park, the river bends around the compressed geologic violence of Mitten Park Fault, an arching ridgeline where we can see eons of exposed rock. We paddle into Whirlpool Canyon, which is darker and narrower than the upstream canyons. The water swirls and funnels as it's compacted. This is where the Echo Park Dam would have been. More than half a century later, we float past a clean and obvious set of drill holes in the cliff wall and a decomposing wooden ladder on the ledge below it.

During the winter, Garth is a math professor at a community college in Colorado, but he thinks the teaching he does here on the river is subtler and more important. People don't really need math, but they need rivers, he says, and he can see how perspectives change after a couple of days on the river, how people soften and open up and start to look around more. For him, being in the canyon starts to feel more real than real life, especially in the heart of summer. "I was here five days ago, and I'll be here again in five days," he says. "And it goes by too fast every time. I'm already getting heartachey, missing the narrower, quieter canyons upstream. And I keep wondering, maybe a little jaded, if it takes expensive, or intensive experiences like this for people to become attached to rivers. Do you have to know a river this well to champion for it?

Camp, on our last night, is at Jones Hole, at the mouth of Island Park. We unload the boats, set up our tents, and hike a few miles up a creek to a panel of thousand-year-old Fremont petroglyphs and pictographs. I stop for a minute alone at the panel, trying to imagine what the river looked like to the people who pecked them out. I sneak off by myself to say thank you to the river. I'm not exactly sure what I believe in, or what I'm trying to do, but like Kerry, I like the idea of rituals.

We rinse one last time in the creek, sweaty from the hike upstream, and return to a final feast around the campfire. We try to make a dent in what's left of the tequila someone bought at the liquor store in Dinosaur, even though we've already polished off the margarita mix. Talk turns to the after: What are you going to remember the most? When might you be back here? I know I've been lucky, I've been on this stretch of river twice in twelve months, but even though I care about it now, I don't know when, if ever, I'll be back.

Oly Lierman, the head river ranger for Dinosaur National Monument, told me she gives out three hundred private and three hundred commercial permits a year. In the 1970s, the rangers developed a river management plan based on user days, and it's stayed the same since then, even though it's hard to get those permits now. The framework for river access, which is the best way to see the park, is narrow.

The recognition of recreation as a viable moneymaker is an unsteady, sinuous curve. It fluctuates, pushed by political threats, pulled back by apathy. Old-timers tell me they've seen activism drop off, and they think fewer young people are taking multi-day river trips, but there's a growing body of support for river preservation, because easy-access recreation has become a lifestyle value. Boat registration in the Uinta Basin is growing twice as fast as the population. More people are paddling because more people are around.[5]

The next morning, the river slows down, the canyon opens up, and for the first time in a few days we can see farther than the next river bend. The Green ignores the rules of geology for most of Lodore, cutting through mountains instead of driving around them. We move out of the Uinta Mountain Group into the softer Weber Sandstone as we get closer to the takeout. Russell, our lead guide, called Split Mountain the skeleton of history. The canyons here slough centuries like dead skin, they're down to their bones. Once the walls drop back, the river feels flat and slow. We start to run into day-trippers who float from Island Park, where there's a road into the canyon, down to the takeout at Split Mountain.

The takeout comes quickly. Suddenly I'm hot and blinking on the boat ramp, dragging dry bags across the sizzling concrete and chucking them into the back of an old school bus. There are numbered parking spots for rafts, and there's a river ranger telling boaters which one to float into. The ramp is crowded with day trips coming in and visitors dipping their feet in the river, and the crush is jarring. It's bustle and crowd after days of just us and the river cutting through the knuckled, sharp-edged rock.

We load the bus and drive out of the monument. I reek, I realize all of a sudden, and I roll down the window to let some sunburned air in. The river bends back around to the west, forming the southern boundary of the monument where it parallels the road. Split Mountain, with its twisted geologic record of the Laramide Orogeny, is to our right. Across the river, on our left, is a big hay farm, owned by a Chinese company that sends its alfalfa across the Pacific. It's stark, coming out of the canyons and seeing bright green fields stretch out.

Coming back is rough every time. By the end of each leg of my trip I am more at home on the river than anywhere else. Back at the O.A.R.S. boathouse, we sort gear, exchange emails, and promise to write, but I've been on enough of these trips to know that these relationships, based in specific experience, rarely hold up off the river. They're deep but fast moving. I go back to the crappy Super 8. When I lie down after my shower, the hotel bed is swaying like a raft.

WE SAVE WHAT WE LOVE AND
WE LOVE WHAT WE KNOW

I was invited on the Gates of Lodore trip because of a man named George Wendt, whom I met on a fishing trip on the glacial and hard-to-reach Middle Fork of the Flathead River in Montana several summers ago. His wife had just died, and our mutual friend who was organizing the trip thought some time on the river might do him good. An older anomaly in our group of paddlers and fishermen, he didn't talk much for two days, and then, slowly, over lunch beers and campfire dinners, he started to spill out some stories. He told us about being huddled in his tent in a 1965 storm on the Yampa when Warm Springs, the biggest rapid on the river, was formed by a mudslide that shot down the canyon, and about his first trip down the Grand Canyon, in 1962, on a raft made of inner tubes and boards.

George was a part of the generation of river runners—the ones who came after the early-day explorers like the Hatches—who turned rafting into a sport and a business. Following the track of Bus Hatch, he capitalized on rafting and cultivated it as a way to get people to care about and save rivers. He knew how powerful time on the water could be. One of his favorite sayings was, "We save what we love, and we love what we know."

He fell in love with desert rivers on a mid-sixties summer break from teaching math in Los Angeles, when he took a trip through Glen Canyon on a Huck Finn–style, strapped-together vessel. He thought the canyon was paradise, and he spent every summer after that paddling and exploring rivers in California and the Southwest. By the early 1970s he'd decided to quit his job and start a rafting company. That company, O.A.R.S, was the first outfitter to run the Grand Canyon commercially without a motor. George branched out to the Yampa, the San Juan, and a range of California rivers, and then started guiding internationally, in Chile and Ethiopia. He fought against the New Melones Dam, which inundated California's Stanislaus River Canyon, because he'd been crushed by the construction of the Glen Canyon Dam, which flooded the canyon that first made him love rivers. He didn't want to lose another place he cared deeply about.

George grew O.A.R.S. into one of the largest rafting companies in the world, but you wouldn't hear that from him. He didn't tell me about any of his early river days until I pried it out of him. On the Flathead he'd dip his cup in the river and drink it straight, something I never do. He seemed like the last bastion of a time when rivers were still wild.

When I told him I was coming to run the Green, he said I should go see Herm Hoops in Jensen, just down the road from Vernal. Herm's worked every job on the river, from guide to shuttle driver to historian for the monument. He's a fixture in the local boating scene, where he's well known for being a conservation advocate and a community builder, as well as a firebrand. He's taken heat from the BLM and oil companies for speaking out against drilling in sensitive areas. He told me later that he learned the even-eyed stare-down he uses in public meetings from Edward Abbey, although Abbey would usually have a .22 by his side.

I call the number George gave me. When Herm picks up, he tells me to stop by whenever and texts me a hand-drawn map to his house, which is also his boat repair shop. When I get there he's finishing cinnamon toast, walking around barefoot in a pair of board shorts and a T-shirt from an extinct raft company. He puts on a pot of coffee, and as it drips, he starts to tell me the story of how he became one of the Green's most fervent advocates.

He came out to Vernal from Vermont in 1966 to run the Yampa for the first time. By 1972, he'd quit his teaching job on the East Coast and moved to Utah, obsessed with the rivers that cut through the Colorado Plateau, the 130,000 square miles of red rock that stretches south from here to the Grand Canyon, He was pulled to the desert by water, and by now every facet of his life has been honed by rivers. His whole house is choked with river ephemera: boat parts, maps, and a first edition of the *National Geographic* that featured Brower's trip to Echo Park.

He flips through a photo album and shows me Anasazi sandals and baskets he's found while poking around up side washes and rarely traveled canyons. He turns to a shot of one of his anniversaries. He's with his wife, Val, sitting on a lip of a canyon. She's wearing heels and a black dress and he's wearing what looks like a marching band conductor's uniform. He says it's what they do every year. They paddle, then hike, into a secret side canyon they've found. They bring candles, and champagne. They dress up. Or they did. This year, because his body is slowing from years of hard living outside, they didn't make it.

Herm is worried that people aren't as tied to the river as they used to be because they don't spend as much time there. He thinks they flow through so fast, if they come at all, that they're missing that sense of exploration that comes from rambling in the canyons. "The difference between then and now is that fees used to be minimal. Hippies could get permits, I would do fifteen trips a year," he says. "It used to be that 30 percent of the cars we shuttled weren't drivable, everyone was a total dirtbag. Now it's all shiny cars and trailers. It's different. The whole GoPro generation and social media has ruined it."

He's become the guy who gets shouted at in BLM hearings and threatened in city council meetings because, to him, being on the river means protecting the river. That's an unpopular position in the gas-rich Uinta Basin, more so when prices are up. He's worried that the river corridor will be lined with oil wells if people who care deeply about it don't stand up to development. And he's worried that no one cares enough anymore. If he feels this jaded, how is anyone else supposed to care, he asks. "If you can't inspire me, how are you going to inspire the other people who don't give a shit? Rationality is not going to win over the average person."

He's not against development, he says, he just wants it done with consideration for the natural resources, especially irreplaceable ones like those Anasazi ruins he's found tucked up in caves. He wants power lines and well pads out of sight of the river, and he wants archaeological sites and petroglyphs protected. He says companies shouldn't be able to drill anywhere in the floodplain. It doesn't sound that drastic to me, but when he brings up any kind of cutbacks to drilling, threats start rolling in. After forty years, he's tired of fighting the same battles over and over again. "My generation, we're a bunch of canards," he says. "I thought in the '60s and '70 we had made headway, but we've turned around and it's all me me money me. We were so close, we valued the water and air and working together as a group. That's probably the biggest failure of my generation, we were so close and failed."

In the next few days, I'm heading into the rarely paddled section between Jensen and the Uintah and Ouray Reservation south of Vernal. This stretch is largely ignored by boaters, so it's hard to find information about it, and it will be my longest spell on the river alone. All I know is that it's flat and remote. The river runs the gut between gas lands and a few lonely ranches, and the USGS map quadrangle of the area is a minimally marked checkerboard. The squares are alternating sections of BLM land and oil fields. When I stop by the local BLM office to ask about camping along the river, I get blank stares, except from the receptionist, who asks me if I'll be paddling with a gun. Herm, who has spent years learning every stretch of the river well enough that he can paddle it without a map, is one of the only people who goes down there with any regularity. I pull out my map and he points out the places where I'll see wellheads from the river, where the BLM has opened up new leases for oil shale drilling, and where I should avoid the land of an antisocial, trigger-happy gold miner. He has a mile-by-mile mental tick list of what I'll pass: granaries, ruins, the beach where he once spent a night half frozen, wrapped in a tarp, in a spring snowstorm. He prints me out notes from journals he's kept on past trips and sends me out with a hug.

After I leave Herm's place, I drive through the adjacent Ashley oil field. The fields are green in his Jensen neighborhood, but when I get away from the river the landscape switches quickly to brown squares drilled and studded with oil tanks. The air is sweet and metallic, and I drive past signs warning about noxious gases.

I head back into Dinosaur National Monument to get a look at Echo

Park from the ground instead of the water. I want to spend some more time in the space Brower fought to protect before I head out of the Vernal area for good and drop down into the part of the desert that's been the most roughed up by energy development. It's wildflower season in the high desert, and the ridgetop road is lined with meadows of lupine and Indian paintbrush. I hike out to Harper's Point, the highest viewpoint in the monument, moving slowly in the oppressive heat, even though I'm hardly wearing anything. I rub lip balm on the cracks in my fingers and toes, which are dried out from being on the water for so long.

This place feels hallowed to me. From the edge, from above, I can see Steamboat Rock, Echo Park, and the Mitten Park Fault stretching horizontally to the east, and the combined Green and Yampa flowing toward Whirlpool Canyon and Island Park to the west, compressing into the narrow gorge. I drive down slippery, steep Echo Park Road to camp by the river. The road changes color with the rocks, white, ochre, buff. Down low, I drop back into the creamy, consolidated Weber Sandstone, smooth tan rocks streaked with oxidized gray. Echo Park is big, wide, and open: orange, then gray, as the light moves through. I walk downstream toward Mitten Park, across the meadow of the open floodplain, then upstream toward the confluence over a slick golden carpet of cheatgrass, invasive but beautiful.

"Once I saw Echo Park I became connected to the rivers of the Colorado Plateau. Somehow there's a great mystery associated with these places," Herm had told me earlier in the day. I have that lit-up and blown-away feeling as I ramble the edge of the river. I wonder how much I'm romanticizing the past, fetishizing the early days of boating the way Herm hunts Anasazi baskets—obsessed with something that seems purer and less contrived.

Herm's decades-long fight against drilling is a piece of preserving the health and sustainability of the Green. So are the fish flows, and Tamara's tamarisk battle, but so is the way Pat is trying to bank natural storage in the Upper Basin, and the way Heather tries to manage the dam. It's abstract and diffuse until you're in it, and then it's even more complex. Now I see that everyone's values hold weight, and that there's a rational argument behind every kind of use. No one is really in the wrong, there's just not enough water for them all to be right. I lie awake in my tent and watch shadows of the moon cross the walls.

The next morning it's eighty degrees by 9:00 a.m., when I drive back toward town. I stop at Betty's Cafe in Jensen for eggs and biscuits, and I pick up a copy of the skinny local paper, the *Vernal Express*. Water, who gets it, and whether there is enough to go around is the undercurrent in multiple stories. There's news about both energy extraction—U.S. Oil Sands, a Canadian company based in Vernal, is trying to open the first oil sands operation in America—and water shortages.

I go back to Vernal and try to call George. He doesn't pick up either his cell or office phone, and both mailboxes are full. In an email he'd told me that he was coming to run the Yampa with his grandson in a few weeks, but he'd also offhandedly mentioned chemotherapy.

I walk down to the O.A.R.S. boathouse a few blocks off the main drag and find Bruce Lavoie, the area manager, in the office, which is spackled with maps of the monument and stickers that say things like "Keep the Yampa Wild." He's smiley with a dark beard and a permanent sunglasses tan. He says that the business can run eighty trips a year between the two rivers, and that last year, they booked almost all of them. "We're doing well. When the economy does well, even if oil prices are down, our numbers track with that," he says. "People in town are starting to see tourism as a value, not as a sore spot."

It's taken sixty years, since Bus Hatch started bringing people down the river, for Vernal to begin to see that value, and it's still not solid. "Our business relies on the preservation of river corridors and wilderness areas, and those places are dwindling. We need them," Bruce says. "We have an opportunity to reach a wide audience, they come from a lot of different places and a lot of different backgrounds. And we're talking about their land. It's a place that's worth fighting for. We're building a group of people who care. The question is how do you leverage that, and when."

Sometimes, when the wind picks up, or my shoulders click and ache, or I'm feeling shaky in my tent on some lonely riverbank, trying to understand the future risk to the river feels like a futile, aimless search. I'd had this idea that I could push myself physically through anything if I was tough and smart and rugged, and that the push would show me something about myself and my place on the river. That being able to do things alone was a sign of strength, not fear. I'd thought I could conquer the landscape and fully understand the problem of water use. But none of that is true. The tough part is connection, looking across lines and knowing when to push the lever on what you think is right.

I'm about to head into the longest solo stretch of my life, but now loss of contact feels harder to bear than its absence. Sometimes I think I'm better at loving places than I am at loving people. Herm knows what he stands to lose. That it's easier to throw your heart into a place, even if, like Herm's, it might get broken eventually. I came into this trip assuming I could be an unbiased observer, but I'm just as entangled in this landscape as anyone. I get that instinct to protect it in the visceral rip of water slipping past. There are parts of this canyon that make my chest ache when I think about them. Echo Park felt familiar before I'd ever paddled through it. Water, for me, and for most people I've met on this trip, isn't something I can think about passively.

Bruce gives me directions to a free campsite on the south side of the river, on BLM land above the Chinese hayfield, where I can look back up into the monument. It's incredibly hot. On the radio, driving in, the weather guy says I can expect only virga—just wind and lightning, no water, all the precipitation evaporates before it can hit the ground. There are well pads in the distance, but I haven't seen another person in miles.

By the time I get my tent set up, the heat pulls back a little and the purples come out in the distant ridges. Split Mountain is a shadow to the north. I start ranging through the scabby hills around my campsite. I'm charged with the idea of exploration that Herm talked about, the idea that you can only really love a place once you submerge yourself in it.

I clamber through crumbling rocks to see what's over the next ridge, joints and flip-flops loose as I skitter down the hillside. The next arroyo is littered with bullet-punched Busch Light cans and a shot-out box TV. It feels both abused and wild, sharp-edged all around. Unloved despite its beauty. Picking through glass shards and fragile crusts of cryptobiotic soil, I am reminded of George's favorite adage, the one that kept him pressing for conservation: we save what we love and we love what we know.

FUTURE RISKS

10,600 CFS

ENERGY AND POWER

It's early morning at the Split Mountain boat ramp, and instead of the crush of commercial trips coming in, it's just me heading out, going downstream. After a week of big rafts, company, and food other than tortillas and oatmeal with O.A.R.S., I'm back in my little boat alone, paddling into the landscape that I've been driving around for the past few weeks. The river is placid and slow moving, featureless aside from sandbars and lethargic eddies. I float out of the monument around the edge of the vivid Chinese hayfield and trace the pronghorn trail where I ran down to the river from my camp the other night.

This nine-day stretch is the longest I'll be alone on the river, and after I pass the highway and the turnoff to Herm's house I'm in unknown territory. From the river, paddling leisurely, I notice things I wouldn't otherwise: a clutch of chairs pulled up under a cottonwood, writing scrawled on the underside of a bridge. The water is a sediment-choked brown, the hills are buff and crumbly, and the thin rim of green around the river is a clear reminder of how narrow the strip of survivability is, how close to the water most things need to be to live.

The majority of the water in the river is already tied up, but there are some major looming consumptive uses that could stretch the remaining resources past the point of sustainability. Climate change is sapping the water, changing how precipitation falls and where while diminishing stream flow. Unadjudicated, and so far unused, tribal water rights have seniority in the basin and could be put into use at any point. And the energy industry, which, as a whole, had shown disregard for the health of the river and the connectedness of the water system, has the ability to buy up water rights when it needs them.

The specter of oil and gas drilling is obvious throughout the basin, and has been since I put on by the Green River Lakes. I've seen pump jacks since Pinedale, but there are deep, invisible impacts, too. We need water to produce every kind of energy, and we need energy to clean, distribute, and treat the water we use. That inextricable tie between water and power is commonly called the water-energy nexus—you can't have one without the other, and that's palpable on the river and in town. Energy development frames every other use around here, and its potential impacts on the river are unclear because of the inconsistencies in how and where companies drill and how they clean up their wells.

The drilling companies that drive the economy around Vernal currently have rights to a small percentage of the Green River's water. But when I start to ask just how much water it takes to drill for oil or frack a well, or how companies get rights, how they track water, and how their use affects the rest of the system, no one seems to be able to give me a straight, conclusive answer. For the most part, as I've bumbled my way through the basin asking about water, people have been willing to talk to me—glad, even, to answer my questions about water rights and fish flows. But when I start to ask around about how much water the oil and gas industry uses, I get shut down or shunted to company PR people, who tell me that it varies so much I won't be able to find hard numbers, or who never return my phone calls.

The industry needs water for different parts of the drilling process: as the base for the fracking fluid, which is water mixed with chemicals and sand or other particles, used to keep cracks in the rock open so oil or gas can flow through; to cool the drill bit; to mix into the cement for the well casing; as coolant for engines; and for dust suppression while heavy machinery tears up newly built roads leading to well pads. Part of the problem is that it's hard to track both the intake of clean water and the output of fracking fluid and other polluted water used in the process, because it's inconsistent. No two wells are the same, and water use—both in quality and quantity—is regulated by a confusing array of federal and state agencies. There's very little regulatory or social incentive within the industry to use water smartly or sustainably. Companies can get water rights from a variety of inconsistent sources, fees for well cleanups are minimal, and if there is a spill or misuse of water, they often pay a negligible fine. As for the output, the EPA keeps a list of more than 900 chemicals reportedly used in fracking fluid, but there's no standard baseline. There's no single source on fracking fluid use or composition, and the Energy Policy Act of 2005 exempted it from regulation under the Clean Air Act, Clean Water Act, Safe Drinking Water Act, and the Superfund act, so no one is officially examining it closely.[1]

As I paddle through, natural gas prices are low and demand for oil has dropped off because outputs from foreign and other domestic sources are high. Vernal feels empty. But even so, maybe because of a cyclical hopefulness, the BLM is still leasing land for drilling. According to the agency's most recent management plans, close to 90 percent of BLM land is open to drilling.[2] Like dam management, oil and gas production is a world built on future expectations. Companies jump to drill when prices are high and ease off when they're low. Towns like Pinedale, Rock Springs, and Vernal cycle through those spikes and dips. Around Vernal, the heads of pump jacks nod along the ridgelines, and I pass the skeleton offices of gas companies along Highway 40. South of town, after I drop out of Dinosaur National Monument and paddle through a checkerboard of alternating BLM land and ranches, I enter the Uinta Basin, home to some of the most productive oil and gas fields in the country, including the Chapita Wells gas field and Monument Butte oil field.[3] The area also holds a fragile but tempting future extractive promise: under the surface of the basin lies the shale of the Green River Formation, which is considered the largest untapped oil reserve in the world.[4]

When I passed through Jackson, Wyoming, on the way to the Green River Lakes, I'd visited Carlin Girard, a water resource specialist who studies how oil development affects fish. He says that once drilling starts, it touches everything around it. Wells and their related infrastructure, like roads, change how water runs off a landscape, what seeps into the soil, and how the land erodes. Once an area is cut up for drilling, it's nearly impossible to reclaim it. But even though he's devoted his career to fish and the environment, he thinks people are the most important thing, and that people need energy. It's naïve to think otherwise. He says maybe we should sacrifice small sections of unloved landscapes to preserve our greater way of living. We should just choose them wisely. No one I talk to about energy development and water use, even the paddlers and the fish people, thinks that we can disentangle the two. We all use oil and gas. But there's an underlying fear that the oil and gas industry's lack of oversight and disregard for the health of the river could upset the balance among all the other interconnected, multifaceted uses. The industry has enough economic power to steamroll other uses, and its track record isn't good.

The Uinta Basin formed as a lake during the Laramide Orogeny 80 million years ago.[5] It's all sedimentary rock, skinny layers of lithified lakebed, which trapped organic matter and formed stratigraphic pockets of hydrocarbons in the heat and pressure of hundreds of centuries. There's been drilling in the Uinta Basin since 1948, when the Equity Oil Company tapped the first commercial oil well in Utah,[6] but because much of the oil and gas is locked up underground in what's called tight reserves—rocks that are hard to permeate—the basin's economy really took off only when fracking became a viable method for accessing those resources in the 1960s. It gained steam in the beginning of the twenty-first century as natural gas began to catch up to coal as the United States' biggest source of fuel for generating electricity. In 2013, 60 percent of Wyoming's natural gas production came from the Green River Basin.[7]

The basin is stippled with wells and crisscrossed with pipelines, but the number of active wells peaks and dips, driven by price. There's no nationwide standard for reporting drilling,[8] but Uinta and Duchesne Counties, which blockend the basin, have been granted the most permits for drilling in Utah in the last decade.[9] In 2012, the Utah Division of Oil, Gas and Mining administered 2,105 well permits; in 2016, they gave out 21.[10] Most of the water use in the industry comes from the drilling and energy extraction processes, but the impacts of the wells carry past their drill dates. Leaky, uncapped wells and unremediated drill sites, of which there are many, also threaten the watershed. It's an issue of quality as well as quantity.

Even though the market is semi-dormant, it's not done. The hydrocarbon-rich basins that the Green carves through have produced oil and gas since the middle of the twentieth century, but the area could be particularly vulnerable in the future because of the large, untapped reserves locked up in geologic formations that historically have been hard, or impossible, to drill because the oil is so entrenched in the rock. The USGS estimates that the Green River Formation contains about 3 trillion barrels of oil, and that half of it will be recoverable, depending on improvements in technology. Those 1.5 trillion barrels are approximately equivalent to the world's proven oil reserves.[11]

Coming off Lodore, where we compared the canyons to cathedrals and were hyperconscious of our wonder, this stretch below the park feels particularly brutal, scabby, and unloved. The changes between the stretches are stark and swift. After the cool, dark canyons, there's almost no shade, and the July heat is stifling. I constantly soak my shirt to try to keep cool, and my skin has a silty crust from the river. It feels cracked in the same way the acrid, sunbaked soil does. My heels split and my nails peel. The water dries me out.

Paddling here is a meditation in solitude, and there's tricky beauty in that, too. I get scared easily when there's no other voice to bounce mine against. But each day I'm calmer. I like making my own decisions, even when they're small, and I like the way the days fold in on themselves, rhythmic, outlined by downriver motion and daylight.

Horizontal fracking wells, the predominant kind here in the basin, use the most water of any kind of well, but how much a single, specific well uses depends on how deep it is, the density of the rock it's drilled through, and what's included in the fracking fluid. "Basically, a modern fracking well is going to take between 4 and 20 acre-feet, which is 6 million gallons of water, in the development phase," Dennis Willis, who worked as a land planner for the BLM in the basin for thirty-five years, told me. That's a wide range, but it reflects the variability in drilling. It's marginal compared with something like agricultural use, but it's a consumptive use, because no water seeps back in to the water table unless something has gone wrong.

The Bureau of Land Management, which manages a checkerboard of land along the river and across most of the region, administers 245 million acres of American public land for multiple uses, including recreation and mining, so it makes the call about which places can be drilled. It was formed in 1946, when the Department of the Interior merged the General Land Office and the Grazing Office. In 1976, the Federal Land Policy and Management Act was passed, effectively ending the Homestead Act and turning all unclaimed land over to the government to be managed by the BLM. How it does so has historically been fraught, especially in places like the Uinta Basin, where drilling has been the dominant land use. "There are people in the BLM who think their job is to make America great by producing hydrocarbons," Dennis tells me. He says he believes in drilling in smart ways, but that the bureau's first job is to protect the resources for the long term. He saw leasing, legislation, and lack of regulation making it easy for companies to come in, drill, and do a shoddy job of cleaning up, all the name of making as much money as possible.

When I'd stopped by the BLM office in Vernal to get maps, I'd asked the woman behind the desk, offhand, what someone would have to do to get an oil and gas lease. She told me it was a mishmash of state and federal domain. Wells here are registered and permitted through the Utah Division of Oil, Gas, and Mining. If a well is on BLM land, the BLM monitors the site through the drilling, production, and reclamation processes. The EPA regulates injection wells and wastewater. Drillers get water rights through the Utah Division of Water Rights, which Andrew Dutson, the assistant regional engineer in Vernal, told me they can access through several channels. Often, they apply for a year-long temporary water right, or they use part of an agricultural water right that's not being used. Sometimes they drill water wells on site, although no place along the Green has significant groundwater.

Sometimes they buy water from cities, or from other water districts, including treated effluent. Sometimes they fill water trucks from fire hydrants and drive the water in.

On the river, storms cycle through in the afternoon, and the clouds scatter pockets of light. As the water level starts to drop, the mosquitoes come up, vicious and bloodthirsty. On the river, in bug season, you can either have shade, up under the cottonwoods, or you can have a lack of bugs, out on the sandbars. I've been choosing the latter, but it comes with oppressive heat until the wind picks up and blows the thunderheads through. Once I'm done paddling for the day, I set up camp, then either lie sticky and naked in my tent, waiting for the evening to cool off, or sit in the silty river, trying to keep my body temperature down.

The cliffs are eroded fingers of red, green, and chalky white. When the storms change the light, some of the clear green comes back up in the river, and it looks alive without the dull brown flatness it gets in the sun. My map shows oil and gas fields all around, and there are wells on the horizon, silhouetted by the thunderclouds and virga. I don't see any people on this whole section, aside from a glimpse of the occasional white oil-field truck on a faraway ridgetop road, but I can constantly hear the din of industry and the whirr of pumps pulling water out of the river. It feels touched and neglected at the same time—spacious and wild, but chopped up.

Different hydrocarbons form at different temperatures and pressures, and the mixed-up geology here holds huge amounts of two of them: bitumen and kerogen. Bitumen is heavy, low-viscosity oil found in oil sands. It's coagulated with water and clay, and to extract it, companies heat the rock and essentially melt the oil out. Oil shale holds kerogen, a waxy substance that can become bitumen, then petroleum, given enough heat and pressure. Releasing sticky bitumen from sandstone, or cooking kerogen to produce oil, requires large, unknown amounts of energy and water, and there are nascent operations trying to tap the basin for both.

On Vernal's main strip, there's an office for U.S. Oil Sands, a Canadian company that has a thirty-two-thousand-acre lease on an area of the basin called PR Springs, where they've spent the past fifteen years working on the first commercial oil sands mine in the U.S. The mine's opening date has been perpetually pushed back because of funding struggles, but when I walk into the cool, spare office, the general manager, engineery-looking American Dan Kline, comes out and walks me through the process they're planning to use. In the lobby there are pieces of oil sands, which have the squishiness of hot summer blacktop, and fake oranges, a prop to explain how they're planning to extract the bitumen using a citrus-based solvent.

Oil sands have been mined in Canada since the 1960s using a water-heavy method developed in the 1920s, called the Clark Hot Water Separation Process, which essentially melts the bitumen out.[12] Dan tells me his company has developed its own solvent, which pulls out the hydrocarbons without using as much water. He shows me the plans for the site, where they'd mine the sands in two-hundred-foot-long strips, then backfill as they go, which means they wouldn't have a tailings pond. They have well rights for water, and he says they'd use a hundred gallons a minute, about one-fifth of a cubic foot per second, to pull out two thousand barrels of oil a day. His sales pitch is good enough that I'm almost sold on it as I head back out into the midday Vernal heat. Back on the street, I remind myself that they haven't started yet—it's all hypothetical at this point.

Even more than oil sands, the biggest potential water suck is oil shale, which hasn't been mined successfully in the U.S. yet. The Green River Formation holds 52 percent of the world's supply, and the Government Accountability Office says its development could have significant impacts on the quality and quantity of water resources.[13] That development could be coming soon. An Estonian company, Enefit, has leased land in the Uinta Basin to mine for oil shale. They're currently in the permitting process, and environmental groups, like the Southern Utah Wilderness Alliance, are up in arms trying to stop them from moving forward. The scope of their operation is still up in the air, but in 2016 they got approval for water rights to build a road, the first step toward setting up a mine.

Around here everyone has a story about some kind of spill or leak.

Tildon tells me what happened to the fish the last time a pipe across the river broke. Herm vividly describes the acridly sweet smell of leaking gas. The water pollution concerns that come with drilling—oil spills, salt leaching out, groundwater contaminated by fracking fluid sent deep into injection wells—aren't just empty threats. The agencies tasked with oversight have historically neglected their cleanup duties. They've ignored uncapped wells and illegal dumping. That adds fuel to the anti-government fire that seems to come from multiple directions because it feels as if the feds are ineffectual. Because of the way industry regulation is currently set up, there's minimal incentive to protect the water source. The well cleanup bonding program is weak, and industries get minor fines or a slap on the wrist if they damage the water source. Companies build their cleanup and remediation plans for new projects on the assumption that the risk of a spill or leak is minimal. It is, for the most part, but it's not zero, and a single spill can be devastating.

"It's all about mitigating risks, but in the process, you realize that the last thing development likes to do is talk about risk. There's a stochastic component you can't predict until it happens. People are like, 'shut up, we're not going to mess up.' And you're like, 'no, you're not,'" Carlin Girard says. "We don't talk about risk in a productive way at all, there's no reward for good behavior, the oil company just pays a fine. It's like insurance premiums, let's reward the people who want to do a good job."

Drilling is a time-sensitive industry because it's market based, and because of that, companies try to bore wells as quickly and efficiently as possible. "I'm not opposed to oil and gas drilling, but there are some places that should be protected, and where it does take place, I'd like to see it take place in a workmanlike manner," Dennis Willis told me. "Typically when a company drills, they subcontract with a driller, then those guys subcontract again. All these fly-by-night subcontracts who are barely making a buck do everything they can to save money and cut corners. Rather than haul a load all the way to the injection well they're dumping produced water into the Green and White Rivers. But I would be surprised if you found anyone to say that. No one is going to tell you that they hire shitty contractors."

Despite the lack of oversight and accountability, there's no pressure on the industry to make change. "It would take some real reform of the regulations, which is always tough to do. Anything that regulates the industry the industry vehemently opposes," Dennis says. "Basically what you're dealing with are classic asymmetrical conflicts. Say there's a smoker in a restaurant. It doesn't bother the smoker a bit, they are more than happy to share the air, and what the hell is the problem with you that you're not able to share. It's a conflict that's both societal and within the department."

Drilling and mining dominate the landscape, but extractive natural resources aren't the only water-heavy energy sources in the area that might impact the river. In 2012, a proposal for Utah's first nuclear power plant, Blue Castle, which would lease 53,600 acre-feet from the surrounding counties on the west side of the river, was approved. The holding company still hasn't raised all of its funds to build the plant, but if and when it does, it has a direct line to the water, which U.S. Fish and Wildlife, rafting company owners, and environmental groups all say could be devastating to their respective priorities. That's a refrain I keep hearing: a serious fear of what could happen, based on what already has. That's part of what I'm learning as I paddle downriver, too, into territory that almost no one sees. There's an appealing suspension of disbelief in focusing only on the current moment. It's easy to do, but that doesn't mean you're prepared for what's coming.

WATER IS WHERE THE FIGHT IS

South of Vernal, the river runs through the Uintah and Ouray Reservation, where the Ute tribe has some of the most senior rights on the Green. Tribal water rights exist in a complicated space both outside of and adjacent to state water rights. Those rights are one of the big, outstanding water uses that remain to be settled, a complicated knot in the string of water rights on the Green, and they bear out all the way down the line. The tribes have some of the earliest prior appropriation dates, so they legally get their water first. But because of the ways tribes have been marginalized by state and federal government, their rights haven't been put to use yet, they weren't taken into account when the Colorado River Compact was brokered, and, because they're federally reserved rights, depending on how the tribes decide to put their water to use, tribal rights could cut off or tighten supply on the rest of the river.

The Uintah and Ouray Reservation is the second largest reservation in the U.S. Its 4.5 million acres stretch diagonally across the northeast corner of Utah, running from the Uinta Mountains to the Colorado border, and encompassing a broad swath of the Uinta Basin. President Lincoln established the reservation in October of 1861, the year before the Homestead Act. The story goes that when Mormon leader Brigham Young was asked about the area, he said that it was useless, so it would be a perfect place for a reservation.[1] From the river, the reservation has the eroded look of long-term unlovability. I see skinny dogs and crumbled buildings in the rare spots where bridges cross the water.

From the road, it's a scattering of double-wides on the edges of rangy hills and low-slung public buildings up on ridges.

I drive into Fort Duchesne, the main town on the reservation, which sits on the crest of a dry ridgeline. I'd been given the name of a tribal secretary to talk to about the tribe's water rights, but she hasn't been returning my calls, so I decide to just show up. When I wander into the light brick and dark wood, '70s-style tribal offices, the security guard tells me I can wait until the secretary is back from lunch. I post up in the lobby under painted portraits of Indian princesses wearing beauty-queen sashes.

There is a tribal government meeting in the auditorium just past the security desk, and people cycle in and out until it becomes busy enough that the crowd blocks the door. I can't see the speaker, but I can hear his slow, clear voice. He's explaining a court case about the dividing line between state and tribal police jurisdiction. It involves a car chase that started on the reservation and continued onto state roads. The tribe wanted full authority on the case, but they were struggling to disentangle themselves from other government jurisdictions that are out of their control. Tribes want sovereignty, but they're physically and legally encapsulated in the U.S. and dependent on the federal government in many ways. They're told to be independent, but they're limited in what they can control.

The murkiness of tribal sovereignty applies to water, just as it does to policing, health care, and education. Indian water rights have been tied up in the political and legal history of the U.S. government's treatment of tribes since the signing of the Constitution, in which they're referred to as foreign nations. In 1851, the United States Congress passed the Indian Appropriations Act, which authorized the creation of the first Indian reservation. It declared tribes outsiders, but still made them adhere to federal laws. For the next eighty years, until the Indian Reorganization Act of 1934,[2] tribes would be nailed to their reservations and forced to eke out a living on the often harsh lands they'd been allotted. That struggle is palpable here. Everything outside of the river is desiccated, and the tribe's baked-in, but unused, rights to water could dictate its viability in the desert.

Bob Anderson, the director of the Native American Law Center at the University of Washington, who advised both the Clinton and Obama administrations about tribal rights, tells me that there's minimal case law setting precedent for those tribal water rights. He says that every water settlement is different and fraught, but that there are three major legal cases that delineate those rights in the U.S., including how much tribes get and how that water is managed. The first is a 1908 case, Winters v. United States, which set the baseline for tribal water rights.[3] When the Gros Ventre and Assiniboine tribes of the Fort Belknap Reservation in Montana won rights to use water from the Milk River for irrigation, they established the Winters Doctrine, which guarantees tribes water rights from the date their reservation was founded. Those rights often pre-date other rights, including those established by the 1922 Colorado River Compact, which gives the tribes seniority. Winters also established tribal water rights as federally reserved rights, controlled by the federal government instead of the states. Unlike other water rights, the tribal rights can't be lost due to lack of beneficial use, which means the tribes are entitled to those very senior water rights, even if they haven't put the water to use yet.

In 1952, the McCarran Amendment pushed administration of tribal rights to the states, even though they're held by the federal government, which diffuses responsibility when they're being sorted out, and which Anderson says adds to the administrative confusion.[4] Ten years later, in 1963, Arizona v. California established a baseline for the amount of water the tribes could claim.[5] As states started to allocate their Colorado River Compact water, they began to worry about those pre-dated tribal water rights. They wanted hard numbers for the amounts the tribes could pull out. The court decided on a metric called practicable irrigable acreage (PIA), based on the amount of water it would take to grow crops on the landmass of each reservation. PIA became the gauge of how much water was enough to sustain the tribal lands into the future.

But the dates and numbers established by those cases go only so far, especially when the rights aren't codified. Bob says that despite that legal precedent, the U.S. government essentially failed to take tribal water rights into account until the 1970s, after many of the major water storage projects in the U.S. were already in motion. Its failure set up a problematic pattern: water rights that came after the tribal rights in priority were developed before them because tribes often didn't have the infrastructure or capital to support big water projects, and because they weren't taken into account when the Bureau of Reclamation was funding big projects. In a lot of places, including here, tribal water rights with older priority dates still haven't been developed, but they can legally pull rank.

"The tribes were underrepresented at compact negotiations and largely ignored as the basins were divided up," Tom Fredericks, a tribal lawyer who has represented the Utes, says. "The Bureau of Reclamation basically didn't do anything on tribal lands when they were building all the big dams in the West." That created a disconnect between the rights the tribes had on paper and their ability to implement water infrastructure, especially where they might have benefited hugely from the help of the federal agencies that were building water projects elsewhere. Water is another way in which tribes have been marginalized and cut off from economic and social development.

Tribal rights are slowly being sorted out. Between 1978 and 2015, thirty-one Indian water settlements were solidified. Most of these settlements required drawn-out and contentious processes, and all of them look different.[6] There are formal criteria for how the settlements are supposed to work, set up in 1991, but Bob says no one really relies on them. "It's kind of a judgment call on when the rights are put to use and it's a real can of worms," he says. "How do the downstream users take that into account? Do you calculate that the tribe isn't going to use it? How do you game it? For instance, the Central Utah Project built other stuff, pulling rank. Now you've got all the big projects built by Reclamation. Phoenix, Arizona, and the Central Valley in California aren't going to disappear." Because those big projects were built without regard for tribal water rights, there's a potential pinch point. When the tribes do try to put their water to use, either they might not have physical water, or they might usurp water from junior users.

I spend an hour sitting on the couch in the tribal office, waiting for lunch to be over, chatting with the security guard. Eventually I get antsy and give up. On my way down the hill, back toward the river, I pass a square white building labeled Water Resources on the side. Frustrated by my fruitless morning, I figure I might as well see what's going on in there. I walk in, awkwardly ask around, and am told to wait for someone to talk to me. I try to join a Wi-Fi network called "water." It doesn't work. Dwayne Moss, a lawyer who represents the tribe on water rights, pokes his head out of his office and says we can talk. He's a sixth-generation Mormon, whose family used to run the biggest sheep operation in Utah, and he has a starchy-shirt tucked-inness that seems to track with that, but he tells me he had a spiritual experience with a Pueblo Apache in 1994, and since then he's been trying to repair what he sees as the damage his ancestors have done to tribes.

In 1965, in the early days of the Central Utah Project, the Ute tribe agreed to give 60,000 acre-feet of water to the Bonneville Unit of the CUP until 2005 on the promise that, as part of the project, a water storage facility, to be called the Uinta and Upalco Project, would be built on the eastern end of the reservation.[7] "As the CUP progressed, the Bonneville project got done, the others got done, but not the Indian projects. By the late 1980s it became clear they would not be," Dwayne says.

In 1980, the state of Utah approved a Ute Indian Water Compact, which was ratified by the tribe in 1988.[8] But the state, tribal, and federal governments all have to sign off on an agreement, and that version of the compact was never approved by Congress, because Utah's congressional delegation still thought the CUP would be completed with the Uinta and Upalco Project. In 1992, when the Central Utah Project Completion Act was enacted, outlining how the CUP would wrap up, it was obvious that the tribal storage project wouldn't be built.

When it became clear that the CUP wouldn't make good on its promise to build the tribal storage project, the tribe declined to sign the 1992 act, because they thought they deserved the water storage rights they'd agreed to. The act still passed without their stamp of approval, and the tribe received some funds in lieu of water. "In '92 they basically settled it. They said 'you won't get those projects' but they gave money into funds for economics, irrigation, and improving the river and flows. That came to $198 million, but they didn't really settle our water rights," Tom Fredericks says. In 2009, a revised version of an agreement to settle the water rights was put forth, but neither the tribe nor the state signed it. Because of conflicting expectations, the water rights have never been settled, which leaves the tribe and the rest of the river in limbo. They can't be prevented from using water, but how much they can use isn't legally delineated.

"The state's position is, 'until there's a water compact with the tribe, we don't recognize it,'" Dwayne says. But since the tribe and the state have never been able to come to agreement, he says the tribe is just going to start developing. "We, right now, for the first time, have developed a water code for the tribe to handle the waters within the boundaries of the reservation."

Bob says he advises tribes to do that, too, because they legally own the water, and because it could get used out from under them if they don't. Even if you have the oldest prior appropriations date, it doesn't matter if there's no water.

"Quantity is not much fought over, it's more bad faith on the state side," Dwayne says. "Our position basically is that the state is still giving out water rights in the basin, even though they know the Utes haven't received theirs. One of the things that irritates me is that in the CUP there was 500,000 cfs of storage rights in Flaming Gorge that turned up when they decided not to build the Uinta Unit. Those water rights were freed up. The Bureau of Reclamation deeded them over to the state, which then divided them out to counties but not the tribe. How can you in good faith do that?"

The residents of the reservation need water to live, but there's more at stake than just survival. Dwayne says that they're acutely conscious of both the economic benefit of the river and its environmental benefit. "Outside the tribe it's always viewed as a commodity, but here on the reservation it's a spiritual matter, too," he says. "River management and protecting a river and the life that it gives has a spiritual nature—it's part of our code."

I leave Dwayne's office and head back down toward the river, which cuts south through the reservation. I stop for one cold tallboy of beer and an ice cream at the Ute Crossing gas station in Fort Duchesne because it's so hot I can't think straight enough to shop for reasonable groceries. The river is dropping, it's past peak, and the midsummer heat feels inescapable, even in my air-conditioned car. I'm exhausted—much more so than if I'd spent the day on the river—by how intractable and unfair the tribes' water fight feels. I inhale the ice cream as I drive, and the beer is lukewarm by the time I get back to my campsite.

At some point, if they don't agree on who gets what, some groups may not get their fair share, and the tribes, which have always been left out of the equation, seem to have the most to lose. Salt Lake City and the urban areas of Utah are slated to keep growing rapidly, and Gene Shawcroft of the CUP says that they plan to develop all the water they divert. They're pulling 102,000 acre-feet from the Green to the Wasatch Front and storing it in Strawberry Reservoir, which was built as part of the Bonneville Unit, allocating the Utes' water rights.[9] The water is already going over the mountains, and you can't cut urban centers off from water, but the Ute tribe doesn't want to get left behind because they were defrauded last decade. The battle goes on, Dwayne tells me, and the tribe won't sign a compact until they get storage rights. "The Ute people are a unique tribe, they're highly independent," he says. "I came with the idea of trying to heal wounds, and it turns out that water is where the fight is."

CLIMATE CHANGE IS WATER CHANGE

I'm alone on the river, with long, wide, midsummer days in front of me, and time slides by slowly. I stop and watch silver flashes of swallow flocks in the pockets beneath undercut cliffs where they make their nests. I read through most of the evenings, slurp oatmeal straight from the packet, stretch out on the sandbars in my underwear—only occasionally unnerved that someone might be watching me. On the first day, as I paddle through Jensen, there's slur-filled graffiti under one of the bridges, which raises my hackles, but after I get away from town, I feel alone, but also, slowly, less anxious.

I think my lack of fear comes from the spareness. I feel better when I can see what's around me. Sometimes at night I wake up to engines firing at odd hours—1:00 a.m., then 4:00 a.m.—and get freaked out that someone might be close, but I never see another person. The landscape here is sere, no shade, just a thin line of tamarisk and coyote willow on the banks. The bluffs erode down toward the river, horizontal ridges of harder rock up high, aprons of softer rock below.

Despite the emptiness, it's never quiet. The range is more stitched up than I thought it would be, and there's evidence that this part of the country has always been used and abused. There are motors running in the distance: water intakes, well pumps, trucks. On the fourth day I hear a dull, uneven hammering from shore, like a cow kicking a door. I creep along the edge of the bank for a mile, listening as it get closer, and finally see the bobbing horse head of a pump jack just above the trees, low enough that there's no way it's not in the floodplain. I assume it shouldn't be here—you can't get a permit to drill a well that close to the river, because it might contaminate the water if it floods—but I guess it's on land that no one cares enough about to check on.

Stories about the West are often stories about people overpowering the landscape around them. They're trying to account for an unknown future, trying to limit their own risk. That's part of why the BLM is still leasing land for oil and gas drilling, even though atmospheric carbon dioxide from fossil fuels is one of the most significant triggers for climate change,[1] the biggest coming unknown risk. Global warming means more variability, more extreme weather events, and eventually less water in the river.

The Bureau of Reclamation, which tends to be conservative in its climate estimates, is predicting that the West will warm five to seven degrees over the twenty-first century.[2] That means areas that now get heavy winter snow, which holds water like a frozen reservoir, will see rain that runs off quickly instead. In the already dry deserts, plants, animals, and people will need more water to survive as the temperatures rise. Climate change means much more variability, too. There will be much wetter years bracketed by longer periods of drought. At the Flaming Gorge Working Group meeting, NOAA forecaster Ashley Nielson said she never would have predicted a spring like this one. Precipitation was 300 percent of their models, and off-the-bell-curve events like that will happen more frequently as the planet continues to warm.

Junior rights holders who depend on the river live in fear of senior rights holders cutting off their water supply if there's not enough to go around. Or that the Colorado River Compact, the fragile interlocking foundation of western water management, might be broken if supply is reduced, throwing the hierarchy of water rights into upheaval. The current infrastructure and the legal and social structures for water management are based on hard numbers and a range of relatively consistent flows, even though that's not actually how the river runs. Global warming is already changing those flows. To survive in the face of that change, and the uncertainty it brings, those hard rules will have to change, too. We either have to become more flexible or more comfortable with risk.

The Green is in the slice of the Colorado River Basin that still has the most water to squeeze out. There's the Ute Tribe's unclaimed share, the unregulated water of the Yampa, and the portion of Wyoming's allocation it's still not using. The Lower Basin's water is essentially all tapped out, but the Upper Basin still has some breathing room, although it might not be that way for long. As that water gets doled out, increased consumptive use in the Upper Basin will weigh on the whole river. The Green is currently the slack in the system. If it's used to its full extent, even though that would be within states' and tribal rights, everyone downstream will feel the impact.

I haven't been on this stretch of river before, so I have no reference point for what the weather should be like, but it has been hot, dry, and not much else since I started. I have a down jacket and a raincoat tucked into my dry bag of clothes, but since I left Fontenelle all I've been wearing is the same pair of board shorts and a gauzy long-sleeved shirt to keep from frying in the sun. There's a scum of sun block and sand on my skin that I can't seem to scratch off. The weather is sucking any moisture from my pores, cracking the sandbars, and causing the river to drop. I can never drink enough, and the midday heat makes me hazy.

Climate scientist Brad Udall grew up in a family that's been paying attention to human impacts on the land for decades. The Udalls are a dynasty in western politics. Brad's father, Mo, was a U.S. representative from Arizona who ran for president, and his uncle Stewart was John F. Kennedy's secretary of the interior, and Brad has taken on a job that's arguably as political as elected office these days. I'd met Brad rafting on the Yampa last summer. It had been a wet spring that year, and the fields of northwestern Colorado were shockingly green when we drove through the ranchland along the Upper Yampa on the way to the river, but downstream, levels in Powell and Mead were dropping. Brad was at work on a study about how climate change could reduce river flow and how that drop might stress rivers as the planet warms up.

By the time I get off the Green, the study is out, and it's scary. Brad and his co-author Jonathan Overpeck found that instream flows in the whole Colorado River Basin could be cut in half by the end of the century because of climate change.[3] Brad says that the projected temperature rise is already well established, so they decided to model future precipitation change, which is less well known. Based on those models, they think that the combination of increased temperature and decreased snowfall will decimate the water supply, especially if we do nothing to substantially reduce greenhouse gas emissions.

To quantify how dry the future might be because of less snow and a warmer climate, they looked at the Colorado River Basin drought in the beginning of the twenty-first century. Between 2000 and 2014, the basin had the lowest average annual flows for any fifteen-year period in the historical record—around 19 percent below average. It was what Brad calls a hot drought, driven by temperature, which means that even though there was only slightly less precipitation, there was much less water in the river. "Most people don't get the idea that the water cycle and the temperature of the earth are joined at the hip," he says. "If you modify the temperature, you modify all impacts of the water cycle."

Brad and Jonathan's models aren't outliers, even though they're overlaying new material. Other more conservative estimates say the river will lose at least 20 percent of its flow by 2100.[4] Even on that low end of the range, that's a huge loss that has to be incorporated into our use of the river's water. "There are tons of papers on the future of the Colorado, almost all of which paint a problematic picture. If you've truly been paying attention, especially in the last two or three years, you should be frightened right now," Brad says.

Even if precipitation goes up a little, which is possible in northern places like the Upper Green, there will be less water in the river and more demand. Peak spring runoff is projected to be twelve days earlier by the end of the century, which means the loss of water storage in snow.[5] More extreme weather events, like decades-long megadroughts, will be more likely. That all breaks down the tenets of a water management system based on hard numbers, consistency, and historical precedent. Of all the potential future changes coming to the river, from pipelines to invasive fish, flow variability due to climate change is going to be the least controllable and most widely impactful.

The effects of a wide-ranging and variable climate add to the implied level of risk in the basin. In May of 2016, Lake Mead was the lowest it had been since it started filling up in the 1930s. The Bureau of Reclamation was starting to get nervous. "The overall Colorado River reservoir system stores four times the annual flow of the river—one of the largest ratios in the world, which is designed to be a buffer against drought, but after years of drought, and increased demands, those reservoirs have gotten low enough that they're hard to refill, especially at the current replacement rate,"[6] it said in a statement about the drought. If low flows and high temperatures continue to deplete reservoir storage and the Upper Basin can't deliver its allocated water, the Lower Basin will reach storage levels that trigger cutbacks. The Upper Basin, which is largely dependent on snowfall and rain, might simply not have enough water.

This desiccated stretch of river I'm paddling now feels starkly different from the dark, deep canyons I was in a few weeks ago. There's water everywhere, but it's barely enough to live on. I didn't bring any fresh water—I'd been filtering river water up to this point—but now the river is turbid enough that it clogs my filter, rendering it useless. I switch to iodine, sucking down bottles filled with silty sludge as soon as the treatment sets in. Everything is caked with the scrim of evaporation. My Chaco tan gets close to permanent. I'd expected my body to change, but it doesn't feel much more muscly or harder, it just feels like I'm slowly eroding, layers scouring off, or dissolving away.

How to deal with water losses is a moral question in addition to a legal one, because there's no real trigger for conservation, aside from fear and faith in the future. Cutting down demand on a shrinking supply while populations increase is an incredibly complicated social math problem. Understanding the cause of climate change and the ensuing ongoing drought, and then actually reacting to it, is critical, but the framework that was put in place last century doesn't account for losses of water from climate change. It all involves a lot of magical thinking.

"We can't assume the current drought will end," Doug Kenney, who works with Brad on the Colorado River Research Group, says. "At some point you have to acknowledge the information exists and you figure out a way to get people to act upon it. There are an awful lot of people who don't want to acknowledge what's already known and what's already been done."

He and Brad both say that the most important big-picture way to deal with reduced flows is to address the root of climate change, greenhouse gas emission. There are two ways to approach that when it comes to water: adaptation—essentially, being flexible around the rising climate and the dropping reservoirs—and mitigation, or reducing the factors that lead to warming. Neither is easy, but adaption is theoretically more practical because it's easier to change human behavior than natural processes.

Adaption is the impetus for all the current structural changes, like the Bureau of Reclamation's interim guidelines, the way ranchers like Pat are building habitat, and allocating water rights for instream flows, but it's difficult to prioritize those practices when the benefit is long term and unclear. Like Albert said in the headwaters, it's hard to want to change when you don't see the outcomes, and because climate change is broad, slow moving, and intangible, it's hard to point to direct causes and effects, at least until it's too late.

There's a disconnect between academic research like Brad Udall's and what's happening in practice because the impacts of climate change are gradual and dispersed. The BLM says that it needs better observational data and modeling tools to build in resiliency as temperatures warm and reservoirs evaporate. Without context or a creeping sense of fear, there's no instigation to respond to the science. "Climate projections combined with drought length has made people in the Lower Basin act, but in the Upper Basin there's been no action around it. There's been some talk and planning, and some of that is driven by the drought, but people still want to build storage," Brad says. Before you even get into adaptation and mitigation, you need to get stakeholders on the same page about what's happening and what they can do.

Brad's study is concrete, far reaching, and frightening, but it's far from the first research to tie water and warming, and so far, that research hasn't pushed much social or political change. People are unlikely to act until they're seeing direct negative impacts. Seventy-five percent of Americans say they care about environmental protection, but they prioritize it below economic issues. It can be easy to push environmental issues aside because they're diffuse and non-immediate.[7] No one feels responsible because the problems seem so massive. "There's also not likely to be any great technological fix that changes how much water is available," Brad says. "The most effective tools in water conservation over the last twenty-five years have been new efficient toilets and higher water pricing. No one's winning Nobel Prizes for those inventions. This isn't high-tech stuff. Likewise, the agricultural sector is moving from flood irrigation to sprinklers and drip irrigation, it's not rocket science. If you're looking for breakthroughs, they're going to be in the form of new incentives that reward people for using less water and give people a reason to try to be innovative." Those incentives are going to be social innovations and pressures instead of technological ones, and those can be harder to apply. Conservation, across the board, is the key to working within a shrinking supply.

An underlying threat of climate change has threaded through almost every conversation I've had about water—people know it's coming, but avoidance as a way to make things manageable starts to feel like a familiar refrain. I think it's why people in Vernal want more drilling. It would make their day-to-day lives seem more secure, even if they might not be in the long run. Albert told me that it's almost impossible to convince farmers to change their practices based on projected climate outcomes because they already deal with so many variables. That's why he's hesitant to adjust, even though he knows shortages are coming.

I'm floating in my own singular kind of avoidance. Life on the river becomes increasingly simple; all I have to do is get myself downstream each day. On the last night, I camp on a wide sandbar in the Ouray National Wildlife Refuge, just upstream from the fish hatchery. It's hot, there's no cross breeze, and I can see the road that cuts through the refuge downstream. I'm not sure if I'm ready to see people after being by myself for so long. I've found a sort of stasis where I can control all the variables, and it feels good. In the morning it's already hot as I pack up reluctantly, pour water into my last oatmeal packet, and push back into the current. The land in the refuge is lower and greener. I see beaver and birds, and make eye contact with a herd of scruffy-looking horses.

Around lunchtime, I see the outline of a bridge downstream, and I start to float into the skeleton town of Ouray, on the Uintah and Ouray Reservation. It's a scattering of shack-like houses and trailers, a former store, some more skinny dogs. My takeout for this stretch. Herm had marked up my map, showing me exactly where I could touch land, warning me that I should be careful about leaving a car there because he'd had trouble with members of the tribe before. People are territorial about their places. Don't listen to anyone if they harass you there, he told me, scribbling around the point of land just before the bridge where it's safe for me to get out. As I float closer, trailing the left bank, looking for my exit, I see two people on shore nervously waving. It's my parents.

FUTURE PLANS

6,820 CFS

THIS LAND IS YOUR LAND

Two back-to-back canyons, Desolation and Gray, split the bluffs of the Tavaputs Plateau from north to south below Ouray. The river cut through the plateau 25 million years ago, as the land was uplifting, and in some places the canyon is nearly a mile deep from rim to river.[1] Fifteen years ago, when I was just starting to think of myself as a river person, I paddled through these canyons on a multi-day kayak trip. This stretch of water is one of the things that got me hooked on the rhythm of river trips in the first place, and I'm curious if I'll see it the same way. At the put-in, I'm trying to remember if the crenellated towers along the river feel familiar. My parents have come out to paddle this part with me. After they pick me up under the bridge in Ouray, they take me back to Vernal for a shower, a burrito, and some grocery shopping, and then we head back to the river together.

From Vernal, we drive through the one-strip gas towns of Roosevelt and Myton, then down a dirt road notorious for ripping tires. The landscape on the drive in is bleak and bleached out, studded with well pads. It's grittier here. And busier, especially once we get to the put-in at Sand Wash. A trip through Deso is one of the best multi-day desert rafting trips in the country: beautiful, remote, and peppered with enough rapids to keep boaters interested. I had struggled to get a permit because it's so popular, and there are two commercial trips rigging alongside us on the boat ramp. You can tell the guides by their garb, just the right amount of beat up and sun faded.

The river is pure chocolate now, thick and creamy with sediment. I am starting to get sentimental. We're inching closer to the end of the Green, and this stretch feels like it encompasses all my questions about water. Everything that depends on the river, from fish to farms, converges in this fragile, threatened canyon. From Sand Wash we'll paddle through public land that could potentially be drilled as part of a contentious land trade. We'll float over the last remaining pike-minnow spawning bar. The BLM manages the river, the commercial rafting permits, and the oil and gas lands on the west bank, while the east side is the Uintah and Ouray Reservation, where access has been contentious in the past. We'll take out in Green River, Utah, next to the Powell Museum, one of the largest archives of river running history in the country. Green River is an agriculture town still trying to eke out a living from the unforgiving land, and the museum backs up to miles of melon fields. On both sides of the river there are pictographs and petroglyphs from the Fremont people who lived here in AD 1000–1150, before drought drove them away. The rock art remains are proof, river ranger Dave Kelly says, that life is hard here in dry times. It is equal parts beautiful and harsh. Or maybe the severity underlines the beauty.

We sort gear and rig the raft, lashing down bags and extra oars, just in case we flip. We strap down a dry box that doubles as a seat for the person rowing, and fasten the cooler to the oar frame. We take turns pumping up the raft. We break a whole carton of eggs in the process. I talk my parents through rigging, and I try to explain to them what I've learned so far and what I've seen over the past two months on the river. Articulating it, after being alone for so long, feels tangled.

My parents have flown out from New England, where property-based riparian rights and a wet climate mean they don't think about water shortages often. The idea of a built-in water deficit is alien, and so is the concept of prior appropriations. After dinner, we sit in the dusk, swatting bugs, and I tell them that, for the most part, everyone is trying hard to use their slice of water in the best way they know how. There are no obvious bad guys or simple answers. We switch on our headlamps, and I tell them I think the question of how to use water in the future feels like it's on a precipice. How do we account for the coming losses and uses? What's the balance between environmental integrity and financial stability? How do we use water more equitably and sustainably while still respecting the past? Can we work within the existing laws and still acknowledge that they are inaccurate and flawed? How do we do better for the future, not just for right now?

Most of the water policy people I've talked to say that there's enough water in the river to go around; it just needs to be managed in smarter, more equitable ways. But that approach contradicts the Doctrine of Prior Appropriations. The current breakdown of allocations is based on history instead of need. To make sure water gets to the areas that need it the most, in the most reasonable, environmentally realistic ways, we need to start thinking of the basin as a whole, instead of hydrologically arbitrary units split up by state.

It's not a new idea. It's actually one Powell considered on his trip through here. He was the one who named this section Desolation Canyon. When they paddled past Sand Wash, he and his crew were sick of being cut off from society and unamused by the sixty rapids between Sand Wash and Green River, Utah. By this point he'd latched onto the idea that land in the West should be split up by watershed, so that water users would know where their water came from and where it was going and could use it smartly within its basin.[2] "This, then, is the proposition I make: that the entire arid region be organized into natural hydrographic districts, each one to be a commonwealth within itself for the purpose of controlling and using the great values which have been pointed out. There are some great rivers where the larger trunks would have to be divided into two or more districts, but the majority would be of the character described. Each such community should possess its own irrigation works; it would have to erect diverting dams, dig canals, and construct reservoirs; and such works would have to be maintained from year to year. The plan is to establish local self-government by hydrographic basins," he wrote in his 1890 report "Institutions for the Arid Lands."

Powell's holistic way of looking at the basin as a unit, instead of splitting up water rights by state, is starting to gain popularity, even though it runs counter to how we currently do things. He predicted that allocating water away from its source would divorce its use from its availability. Users wouldn't have context for drought if they didn't know exactly where their water was coming from, and it would be easy to overuse. He was right. Currently, because water is a collective resource, and because the burden of overuse isn't borne evenly, there's no compelling incentive to conserve, aside from forethought or guilt. I see that myself every day in the shower at home, letting the tap run. That's slowly starting to change as drought-based scarcity prompts fear, and as research predicts it will get worse.

In the morning we make eggs and bacon, dancing to avoid the mosquitoes while we finish rigging the boat. We top off the raft tubes, check our straps, and push off into the languid river, trying to get ahead of the commercial trips so we can have first dibs on campsites downriver. We take turns rowing through the flatwater, adjusting the oars so that the blades are straight and the handles are almost touching. I show my parents how to settle into the middle compartment, butt on the dry box, feet against the crossbar, and give them an abbreviated rapid-running lesson so I don't have to row for the next six days straight. The river slowly pushes us around corners, past tamarisk-choked banks and flat, sandy river bottoms where streams run in during the spring. I pull out a map that Herm gave me, annotated with his favorite secret spots, or at least the ones he'd been willing share.

Herm is particularly worried about this stretch of river because he thinks it's both the most special and the most vulnerable. The BLM manages the area, so it's more susceptible to drilling and other extractive use than some of the other high-value recreation areas, like Lodore, which is managed by the Park Service. Part of the area around Desolation is a Wilderness Study Area, and stretches of the river have been deemed suitable for Wild and Scenic designation, but none of that has been officially sanctioned, so its future is tenuous, especially because there are proven oil and gas reserves on the north side of Desolation Canyon.

Before I'd left his house a few weeks back, Herm showed me a letter he'd sent to then-President Obama speaking out against a proposed land trade bill that would open up parts of the area around the river to oil and gas drilling. In it, he argued for balance and judicious leasing for drilling. "The question is not whether we need oil and gas," the letter read. "The question is whether we need that little bit of oil and gas INSTEAD of those beautiful, quiet, pristine places to get away from our daily pressures. When the land is protected by legislation those oil and gas resources will not disappear. Indeed they will be more valuable in the future should the national need to use them arise. If we take care of what we have and protect it, the extractive industry will thrive and the outdoor industry will continue to thrive."

That seems reasonable, and even smart, to me, but the threat of development is ever present because so much of the BLM land in the area is open for leasing. Before we launched at Sand Wash, we stood in the shade, scratching at mosquito bites, as Dave Kelly gave us a ranger talk. His spiel was low on safety and logistics, high on the fact that public land belongs to all of us. If you divided it up by population, each U.S. citizen would get 30 acres, he said, and maybe yours, emotionally, is right here in Desolation Canyon.

The night before, as we rigged the boats, he wandered down from the ranger station to chat with us. When I asked him what he thought Deso's future looked like, he said he's sometimes frustrated by his employer, the BLM, and by the uncoordinated mismanagement of public land, because of the way it might impact water quality and how much water is used in the basin. He's down here all the time—his work weeks alternate between paddling the river and staying at Sand Wash—so he hears different takes on the future of the canyon. He wants to see the river preserved, but he knows that energy development is a real priority.

He knows that the area around Deso could easily be leased or sold off, and that any time land use changes it significantly changes the river. Up high in the basin, when ranchland is converted to urban neighborhoods, more water is used consumptively. That means less water is available farther downstream, cutting the slack in the system, even if the upstream users don't necessarily have rights to more water. Down in these rangy canyons, land use could be changing in a large-scale way.

There's a habitually reintroduced land trade bill, spurred by a growing movement against federal land management,[3] that would threaten the land around the Green by transferring public land to state or local governments, which have historically sold off or opened it. It's been brewing since the 1980s, when so-called Sagebrush Rebels fought, sometimes violently, against federal public land designation after the Federal Land Policy and Management Act passed in 1976, ending homesteading. The movement reignited in the 2010s around groups like the anti-federal-government Bundy family, who staged a standoff on a wildlife refuge in Oregon. The bill gained steam under the Trump administration, with its lax environmental protection regulations and heavy pressure from energy interests, and it's a threat to public land and water across the West.

Ray Bloxham, who is the field director for the Southern Utah Wilderness Alliance, a nonprofit dedicated to desert landscape preservation, says the river will be threatened by oil and gas development along the northern portion of Desolation Canyon, above the Wilderness Study Area boundary. "We're starting to see development on sections on the rims, you can hear it from the river on quiet nights," he says. SUWA's goal is to preserve as much of the untouched desert as possible. They push policy, and they go through every drilling lease in the area to make sure it's valid and up to date. It's tedious, exhausting work, but he thinks it's crucial, especially in the face of an antienvironment administration. There are so few truly wild landscapes left, and he says he thinks Deso is one of the last remaining good ones. Ray says SUWA sees policy and federal land designation as the best way to protect sensitive resources like the river, especially because water rights for instream flows are so hard to get. "The best strategy we have is the legal department. If you look at the long run, it's political pressure. We would love to get a wilderness bill, that's the proactive approach to protect these places."

Other advocacy groups, such as American Rivers, are trying to protect the area through political action as well, but their voices, especially here in Utah, are often drowned out by energy industry lobbyists and local distrust of the federal government. They've spent the last decade trying to get parts of the Green designated as a Wild and Scenic River, which would give it federally reserved water rights, but that designation hasn't moved far because opponents see it as overly restrictive. "It's some of the most high-value, wild, iconic river reaches we have in this country," American Rivers' Matt Rice says about the Green. "But it's been really, really difficult to get any momentum around it."

The canyons dip and curve, forming amphitheaters, hoodoos, and spires. The walls are striated, striped tan, green-gray, and very black. They funnel the wind, which picks up in the afternoon. We fight it through the day's last few sections of rapids, constantly getting pushed upstream, until we find a flat, protected beach. The light starts to take on a low evening slant, which comes slowly this time of year. We swim a little, then boil water for pasta and talk about whether or not the world is more messed up now than it ever has been before. As we sit sipping our beers, a beaver swims up the eddy, sighing and moaning. When it flicks its tail and dives under water it sounds human.

This morning, Dave sent us off with the reminder that we're lucky to be here and that there's something rare and special about getting the chance know a place like this so well. We are all intruders, but spending time here is a way to be connected to the landscape, to know its worth beyond extraction. And, as recreation becomes more of an economic driver, he says it's even more important to hold onto boundaries, to keep the wild places wild. Preserving land, or using it smartly, even in places where water rights are divorced from property rights, can help us hang onto water and keep it clean.

The canyon gets steeper, stonier, and more layered-looking as we go deeper. The rock is younger and sharper than Lodore's, but the walls are still high, oxidized black, buff, and iron-rich red. River miles wind and travel differently than other miles. You can row a little harder, but you can't do much else. It feels like we are skimming through the canyon too fast, temporal and impermanent. We stop to hike to petroglyphs that Herm has marked on his map. On the rock shadows we pick out scratches and swirls in the ledge. I wonder how dry it would have to be for you to abandon your home.

YOU CAN'T JUST SELL
OUT TO A CITY

It's hazier in the morning, which brings out the contrasting color in the cliffs. Deso is deeper than the Grand Canyon here, the intertongue-ing rock layers of the Wasatch and Green River Formations climb five thousand feet to the rim of the plateau, and this stretch of the canyon is younger than almost anywhere else along the river; it's raw edged.

We stop for water at an abandoned homestead, Rock Creek Ranch. In the early 1900s, the three Seamount brothers squatted on the land, and over the next few decades they ran cattle, build solid stone houses, and planted fruit trees, which still line the clear creek that runs into the river above a curving point bar. The Seamounts left in the 1920s and sold the farm to another family, the Jensens, who stuck it out until the 1960s before giving up on ranching in the remote canyon.[1]

We refill our bottles in the low water of the creek, trying not to muddy it as we wade in, then wander through the ruins. Inside the stone buildings it looks as if the Jensens left in a hurry, assuming they'd be back. There are cracking, dust-filled leather boots kicked aside, jars on the wooden shelves, posters curling on the walls. There's evidence of how hard they tried to make it work—there's shade under the fruit trees, and the chinks of the buildings are still snug, almost sixty years later—but a few dry years can flip the balance of viability for a farm or ranch.

In the balance of future water use, agriculture is often assumed to be the sector surplus will come from. On a broad scale, it makes sense to shift water from agriculture to other uses because agriculture uses 80 percent of the supply, but on a granular scale, shifting the balance would mean slaughtering some people's way of life, or changing the way water rights are allocated. It's currently hard to slice a little water out of a right without sacrificing the entire thing, and water rights are so closely guarded that no one gives them up without a reason. Different uses rub up against one another, and fear doesn't really allow for generosity. So the current question is, if we're going to have to reframe the way we share water, what will drive the reshuffling?

"You have to do good for rural areas, you can't just sell out to a city. It helps in terms of the macro scale, but if you go to a rural area and have farmers sell their water, then pretty much every job in the area is gone," Doug Kenney says. Stripping people and ecosystems of water they've depended on for decades is a problem socially, morally, and politically, because livelihoods are so directly tied to water. And it hasn't gone very well where it's been done in the past. The practice of "buy and dry," in which growing cities buy up water rights from agricultural areas and transfer the water off the fields, has scoured rural economies and led young people from farming families to move to urban areas. People like Pat O'Toole worry that the practice rips the bottom out of the food system. He says you can't grow food if there's no water, or people to farm.

Because of prior appropriations, you can't just pull a little bit of water out of agriculture. To get farmers to give up some of their water rights, they have to see the value in changing their ways. Figuring out flexibility in the face of shortages requires a reworking of the current rigid compact. And that flexibility can be scary for farmers and ranchers. They're often hesitant to let even a little bit of water go, even temporarily, because they're worried they won't be able to get it back if they need it in dry years.[2]

Environmental groups, along with state and local governments, are working on programs to reallocate water within the bounds of the compact. There are currently two potential channels for change: markets, which let agricultural users lease or trade their water rights without sacrificing them in the long term, and changing government regulations, which are driven less by supply and demand, but which protect historical rural values and water rights.

To create a market for resources without direct monetary value, or for those that haven't had it in the past, like recreation or flows for fish, there has to be an implied value. Markets have to exist within the social and legal context of water use. They have to come with safeguards to make sure people don't sell more than their allocation and that they don't injure downstream users. And they have to make sense for both sellers and buyers.

There are several experimental market models in different stages of development across the West. Most of them incorporate temporary transfers of water from the agricultural side, such as water leasing or intermittent field fallowing, with compensation from non-agricultural users, like cities or environmental groups. When I floated through the flood-irrigated fields below the Wind River Range, I was in the midst of one of them.

On the Upper Green, there's a market for ecosystem services called the Conservation Exchange[3] in which ranchers can receive payment for land stewardship. It's a partnership between ranchers in the area, like Albert, and the University of Wyoming, the Sublette County Conservation District, The Nature Conservancy, the Environmental Defense Fund, and the Wyoming Stock Growers Association. The program started in 2014 with a market for sage grouse habitat. Ranchers who protected breeding and nesting habitats for the birds could apply for credits. Energy companies, which had to pay into conservation funds for land disruption when they drilled because of a U.S. Fish and Wildlife program to protect the birds, could buy those credits. They offset the damage done by their construction by paying ranchers to protect other places. Based on the success of the sage grouse program, the Conservation Exchange is working on setting up a similar market for water. The idea is that ranchers would use slightly less water, send the surplus downstream for other uses, and be compensated for the losses they might accrue by irrigating less.

Water is trickier because it's not easy to guarantee that an exchange has value to both sides. The sage grouse program had clear regulatory boundaries because energy companies were required to mitigate for their construction. It made sense for the industry and for the ranchers, who profited when they set aside habitat. With water, there's not as much of a clear-cut need on the buyer's side, but Kristi Hansen, a water resource economics professor at the University of Wyoming who helped set up the exchange, says that she thinks there will be willing participants as the pressure on water increases downstream. They're working on building ways to quantify the value of a healthy water system.

While the Conservation Exchange is slowly finding its footing in Wyoming, the most widespread model for a water market system downstream is water transfer, in which users lease their agricultural water rights to water districts or cities for a fixed amount of time. They don't lose those rights, which is a shift from the historical precedent. Aaron Derwingson, agricultural coordinator for The Nature Conservancy's Colorado River Program, says that they've set up trades in rural Colorado by paying ranchers and farmers to fallow their fields and keep water in the streams. He says it was important to start with private landowners and to set up a system that worked for locals. He doesn't think change can start from top-down government policy. The market is currently bankrolled with funds from the Colorado Water Conservation Board, grants from the state, and philanthropic funding. Aaron thinks it can be self-sustaining eventually, but there are challenges. There's a cap on supply, and there's a necessary level of trust that can be hard to establish on a large scale. "If it doesn't work for one of those partners it isn't going to get there. We're realizing that as much as we want it to be a political thing, it's really a social thing," he says. "There's still no shortage of paranoia about water and water rights. It's promising to see some individual shifts but there's still a way to go. Unanswered questions are the name of the game."

The Seamounts and the Jensens weren't the only people who tried to eke out a living in the canyon. Bootleggers hid liquor stills high in the cliff layers and made moonshine out of the farmers' leftover peaches. We climb up to some of the caves where they distilled and see the detritus of hideout life: broken glasses and abandoned camps. Downstream, just below the curving wave train of Cow Swim rapid, we stop on the left side of the river at the McPherson Ranch. Jim McPherson started homesteading here in 1889.[4] He built a beautiful, tightly jointed stone house and fenced off cattle in corrals pinned up against the cliffs. He passed the property down to his daughter and her husband, but because it was on the east side of the river, the family ranch was transferred to the Uintah and Ouray Reservation as part of the land taking for the reservation by eminent domain in the 1940s.

The Utes tried unsuccessfully to run the property as a riverside motel, and the bones of that venture sit next to the skeleton of the McPherson Ranch. Abandoned several times over, and now slightly menacing, it's another unintentional memorial to the hardness of life in the desert. There's a blocky '70s-style building, with a broken Coke machine and busted urinals still inside, next to McPherson's carefully built stone buildings. We walk through the ruins, and I get the ghostly jumpiness of being watched. There's a layer of sediment on the broken furniture and cracked glass in the remaining windows, but it feels like the people who lived here left expecting to come back. We have lunch on the beach, then paddle out into the glaring sun as the constant afternoon wind picks up.

The wind is constantly stiff and up canyon, the kind that pins you in an eddy and keeps you there, compressing sand and scouring our faces. The thermals howl so hard that even when I spin and row the boat backward, pulling with every muscle, I still get pushed upstream. By the time we inch down to a scrubby beach and decide that it will be good enough for the night, I am ragged, crabby, and run thin from fighting wind and getting nowhere. I feel eroded. At camp, the sand has a shell of breakable wind crust, and the river reflects back the canyon's colors: red as the buttes catch the light, then the dusky gray-green of the tamarisk and the muddy beaches.

We've settled into a tempo of family river days, but tonight I'm discouraged and resistant. I'd wanted to show my parents how capable I'd become, and how this river trip had given me poise and patience I hadn't had before, but instead I eat Pringles and pout while my mom cooks me dinner, trying to remember that vulnerability isn't the same thing as weakness.

McPherson came to the canyon on a mission to prove his worth, and to turn the rugged landscape into something tamable. I love that myth of the frontier—of going to wild, lonely places as a way understand yourself and to test your toughness—but much of it is just that: a myth. The rain never followed the plow. It's impossible to control a desert canyon by willpower alone, and doing risky things on your own doesn't make you better or tougher.

That's true for water, too, which is still managed, in many ways, on the myth of western isolationism. But doing everything on your own is antithetical to change. Aaron is working to make small-scale, trade-based, local water exchanges more appealing, but the other line of thinking about reallocating water is that change will have to come from broader government regulation.

There are active water banks, which allow for the temporary transfer of water rights, in some western states, like Washington and California,[5] but none of consequence along the Green. In 2016, Anne Castle, the former assistant secretary for water and science at the Department of the Interior, advocated for a statewide Colorado water bank and for new legislation to allow for a "sanctioned, easy to use, profitable yet protective mechanism to enable transfers of both direct flow and stored agricultural water rights." Anne says that a bank, in which users can deposit their water rights for a specific amount of time instead of giving them up forever, builds in flexibility that benefits everyone. It's a good concept, but a bank needs an overseeing body to make sure it's fair, which is why she thinks government supervision is important. It also needs interested depositors and borrowers. There needs to be a baseline of available water so that buyers know they'll have a resource. Because cities want to guarantee their residents a reliable amount of water, they're reluctant to stake their water supply on borrowing from a water bank. That's the problem with leasing, too; everyone wants their rights to be secure.

It's hard to create both stability and flexibility and to do so in a way that considers the most vulnerable users, or the environmental downsides. That's why buy and dry has been a problem in the past, and why people are skeptical of letting go of their water rights, even temporarily. Water is currency, and in places like the dried-out town of Green

River, Utah, where we come out of Deso, it's clear that lack of water is a poverty problem exposed by drought.

Later, in town, which is a grid of fully abandoned houses and yards lined with cracked trailers and broken furniture, I buy two tiny honeydews from an older woman at Vetere's melon stand. She tells me they're ripe when the skin is a little sandpapery; the one I've chosen is too smooth, and she grabs me another one. Her skin is wrinkled and sandpapery, too. Green River is known for those melons—they're one of the town's biggest economic drivers, and Vetere's is one of the major operations in town—but I've spent less than four dollars.

If conditions get drier, or if cities are willing to pay huge amounts of money for imported water, we might reach a point where it won't be viable to run cattle on the Upper Green or to flood-irrigate fields in the high desert. It might not make sense to grow things in Green River for just a few bucks of profit, any more than it did in the canyon for the McPhersons or the Jensens. Water banks would ideally build in flexibility and options so that people like the Veteres can decide for themselves. Foresight doesn't necessarily have to be at odds with what's currently sustaining people, but things will have to change to make that true.

"We know all sorts of ways to do better, you just have to give people incentives," Doug says. "Last century the solution to water problems was technology. I think that's just in the backseat now. Now the solutions are incentives around using less water."

GETTING COMFORTABLE WITH RISK

In the canyon, Wire Fence and Three Fords rapids come one after another, with hardly enough beach space in between to collect yourself after the first wave train. The shore is thick with tamarisk, so we hike a deserty overgrown ridge to scout them both. As we stand on the shore watching water pile up, my stomach drops. Both rapids are channelized and rock studded. The margin for error at Wire Fence is narrower, but Three Fords pushes hard against a rock wall on the right side so that you have to fight the current to avoid getting pinned. I hike back upstream silent, visualizing the markers I've picked out for myself, trying to run through the lines in my head, but the memories have almost faded by the time I get back to the boat.

We push off, and I pull against the persistent up-canyon wind, fighting to line myself up for the channel. Rowing a heavy oar boat takes both patience and fast-twitch, spilt-second reflexes. You have to set your angle early and let the river take you—you can't fight it all the time—but you also have to be ready to react, and know when to ship your oar and when to pry against the current.

Most of the technical rapids in Deso are class III, depending on the water level. The river compresses into wave trains and pours into holes, bending around eroding corners of the canyon. Nothing too scary, just enough to make you constantly vigilant, especially with the burden of never having seen it before. I don't know the strength of the holes, or whether they're powerful enough to flip a boat.

We hit the tongue just left of a marker rock in the middle of the channel, and I push hard to the center to avoid the diagonal wave echoing off the right shore. I punch the wave, spinning to hit it straight on. The boat sucks back a bit, then pushes free. Suddenly I'm rowing for the middle of the channel, clean.

The river bellies around a bend to the right, then drives back left, building into the compression waves of Three Fords. In the slack water above the rapid I fight the urge to recalibrate. I'm looking for my markers at the top of the rapid. I nose just right of the guard rock, exactly where I want to be, then jam my oar against the rock, throwing it out of my hand, and pushing us off line. I regrab the oar and straighten out in time to hit the second hole head on. From there it's just rock dodging, the kind of river running that feels like a video game. I get a "Good job, kiddo" from my dad, who never calls me kiddo. We float on, into the mouth of Gray Canyon, past the last pikeminnow spawning bar. So much of paddling whitewater is calculating risk: planning for what you can't see by reading the bubble line leading over the edge of the rapid, trying to commit the dangerous parts to memory so that you don't get sucked into them, and trying not to hurt other people along the way.

Water management is risk management, too. "Throughout the whole last century, if you needed more water it always worked out somehow, but it doesn't work when you get to the point where you're storing every last drop," Doug Kenney says. "You have to talk people through, and explain that for every new reservoir you try and fill you're putting more stress on the other parts of the system. Things are changing and we should behave in a way that limits our risk." I'm starting to think that the undercurrent of risk is an important driver; it galvanizes people. We're going to have to grapple with how much risk is reasonable, and who bears the burden of it, when climate change cuts into supply and demands grow. Reducing risk along the whole river requires changing the conversation about what individual users think they deserve and the legal structure around how they get it.

The closer I get to the confluence, the more I think about the river as a whole, and about how to smartly shift the balance around. Brad Udall says that if he could completely flatten the water rights system and build it back up, he would put critical human needs first, the environment second, all other uses in a third tier. It would be a total re-structuring, but he thinks it might start to happen on its own if things get desperate enough. "There are some really obvious, highly valuable uses that need to be taken care of first. The economic powerhouses will get that water. You're not going to grow alfalfa and not get water to Denver," he says.

We need to think about drought contingency planning like Powell did, on a basin-sized scale, and about balancing supply and demand in a way that respects prior appropriations but also acknowledges that eventually there won't be enough water to fulfill all the obligations. "We need to accept the fact that Lake Powell and Lake Mead will probably never be full again. We need to deal with the reality that not everyone will get as much water as they want or as much as they were promised. We just have to use water more efficiently, a little belt tightening and that sort of thing, and then most of this becomes very manageable," Doug Kenney says.

Because supply works in one direction—downriver—increasing consumptive use higher up in the basin stresses everything downstream. To account for that in the future, water managers are working to come up with contingency plans stating what will happen if there isn't enough inflow for the Upper Basin to meet its obligation to the Lower Basin, or for either of them to meet their obligation to Mexico. They're talking about voluntary cutbacks, or reservoir levels that trigger mandatory cutbacks. It's all connected, so it needs to operate that way.

If we moved toward managing the basin as a connected system instead of state by state, water could be moved around to where it's needed most. The reservoirs, from Fontenelle down, are a bank of stopgaps, and recently there's been talk of using them together to shuttle water downstream. When Lake Powell gets scarily low, the contingency plan is to push a big pulse out of the upstream reservoirs to preserve minimum power generation and meet obligations at Lees Ferry so that the Lower Basin gets its share.[1] But that's not a supply plan that will work often. Brad Udall says it would probably be a one-time gamble. Those reservoirs have limited capacity—they're small compared with Powell—so they're not silver bullets. Demands in the Upper Basin will have to be managed to be able to augment supply.

If slack is built into the system by moving stored upstream water down from the lower-value irrigated grasslands in the Upper Basin, it would carry the politically polarizing consequence of limiting supply on the Green. Ranchers high in the basin might not get water when they need it. That's the fear people in the Upper Basin have: that the big rich downstream cities are coming for their water.

In 2007, as drought crept through the Southwest and stayed, and Powell and Mead both hovered around half full, tensions rose among the states. Conflict began to feel unavoidable as supply shrank. To address that fear, and to plan for coming shortages, the Department of the Interior developed interim guidelines for the two major reservoirs. In the agreed-upon guidelines, which apply until 2025, there are three Lake Mead levels that trigger shortages in the Lower Basin. At the first level, total water use is cut by 4.4 percent in the three Lower Basin states. But even that doesn't shake out equitably: Arizona's supply gets cut by 11 percent and Nevada's by 4 percent. California's allocation isn't reduced in the first round of cuts, because of interstate treaties with Arizona, but it would be cut at lower levels. Lake Mead hasn't hit the first level yet, although it's been close, and as it starts to drop, politicians in the Lower Basin, including in California, have said that they would be willing to take voluntary cuts beyond those in the guidelines. The politics are going to get much more complicated, but it's groundbreaking that those states have agreed to be more flexible, even if they're not legally obligated to do so. It's a signal of change in attitude.

While we're scouting Coal Creek Rapid, a group we'd passed the day before comes floating through. I'd noticed their eight shiny Sotar rafts and plethora of gear on the beach. They don't stop to scout. The lead boatman stands up in the oarlocks, looks downstream, then makes a move left to run the meat of the wave train. We watch the rest of the boats paddle through, following his line, then we get back in ours and trail close behind them.

We catch up when they stop for lunch. They're a group of friends from Oregon, Michigan, and Montana. Their average age is pushing sixty and they say they do a trip like this every year, ticking through the major multi-day river trips in the U.S.: the Rogue, the Jarbridge Bruno, the Snake. Herm had told me he hardly ever sees young people doing multi-day tips anymore, and judging by the crowd we've seen, I'd guess that's true. He's worried that there's no new blood and that lack of time on the river can make it hard to have a stake in it and a targeted sense of activism. He's worried that the people who care deeply are getting too old to paddle.

I don't know if you can truly love a place until it's beaten you around a little and shown you all its sides. You don't necessarily have to commit yourself to the canyon forever to care about the river and want to protect it, but I think I needed this attachment to both people and place that I got after a few cycles of constant contact: some wind battering and some broken eggs. The distance and discomfort of my journey made the threats of climate change and overdevelopment real. I got to the point where I couldn't look away. Edward Abbey, who spent a lot of time in these canyons, wrote that part of love is anger. I haven't met anyone on the trip who works with water who came to it passively, or from obligation, not even the government wonks. Everyone knows what they stand to lose. I hadn't realized that until I came here and stayed. The river cracked my heels and changed my mind. Here, almost at the end, I feel less scared, but more vulnerable because I realize how tenuous the connections are. How thin the river will have to be spread and how many tricky compromises will have to be made.

Demand management by way of conservation is the most tangible solution. Even if we shuffle water and build more reservoirs, the supply won't increase significantly, so the burden falls on cutting desire. "Consumption is down in most places, most cities at least, and even the fastest-growing cities are at least stable. That's tremendously useful," Doug says. "It's not terribly sexy, but conservation is the answer to so many of these questions. The trick is to incentivize it."

Those incentives come from showing people what they stand to lose, and why other people's losses might affect them. It's about balance. Conservation can look like a lot of different things: water markets, leveling fields to maximize crop absorption, cutting energy use in cities, finding ways to reduce reservoir evaporation. It's getting people on the same page, building in flexibility, being vigilant in not abusing the existing supply. That will take reshuffling water rights, shuttling water to cities, expanding the supply to incorporate water for recreation, tribes, and fish. It will take breaking down history and myth and hanging onto the bits that are important. But that's a tough proposition. The river cuts through different priorities, and culture is hard to communicate, especially when it's deeply ingrained. Overuse often comes from fear. We all look through the lens of our own experiences. I saw that when I went to the rodeo with Will, and I saw it in Vernal with George Burnett, when roughnecks came up and hugged him on the street.

Culture bleeds into the question of what we're comfortable risking—for ourselves and for others—and what we prioritize. It's a question most people, myself included, don't consciously ask when we use water, but one I think we'll have to ask soon. Not addressing shortages and ignoring risk is a very real way to put people in danger. As I float closer to the confluence, I think a lot about a question Carlin Girard asked me in Jackson, before I even put on the river: Do we sacrifice some things and some places for a greater good? And if we do, how do we agree on that?

I don't think there is one specific answer or one singular sacrifice. It's a mesh of adaptive management and building in flexible structures that don't screw anyone over, or make anyone feel screwed over. It's more stringent regulation for industries that damage ecosystems, and it's a matter of building predicted climate change into every part of how we operate. The places that made me the most hopeful were the ones where people were trying to get on the same page. Everyone works within their own reality, and sometimes multiple realities can be true. "That's the challenge for this whole basin," Matt Rice from American Rivers tells me. "There aren't good guys and bad guys."

I know that it's hard to hold all of that in balance. It's easy to just think about your section of river. I have to fight that. For me, the canyon becomes insular in a way that can be misleading.

I turn on my phone at the takeout and get an email that says George

Wendt has died. Complications caused by non-Hodgkin lymphoma, on July 9. We drive back to town through alkali flats where it looks like the crusted ground can't hold onto anything other than itself. Nothing grows that isn't irrigated. My parents take off, and I'm gutted and lonely in Green River, Utah, reading George's obituary in a scabby-looking coffee shop in the empty town.

I still have a list of questions I'd planned to ask him scrawled in my notebook. The big-picture interview-y ones we never got to talk about:

- What was it like when you first came to the Green?
- How has everything changed?
- How can you use rafting to protect and preserve wild places?

I loop those questions back in my head. They're more poignant now that I'm not just grilling him about the history of the company: How do you make it work? What have you seen that gives you hope?

George had made a mission of getting people out on the water to give them common ground. He was part David Brower, part Bus Hatch, primarily good intentions. And that, from what I can tell, is what it's going to take to create consensus. It's the principle behind the Upper Green Conservation Exchange and the Flaming Gorge Working Group. It's finding ways to make conservation a practical, understandable, appealing part of everyday life, and making sure it's not at odds with the economy. It's not a novel strategy, but it's the best one I've seen to change the culture around water. George's generation melded business, recreation, and conservation, and now we have to take it further, to bring in agriculture and urban planning, to hold all those uses in some semblance of balance, to show the other side. As George always said, "We save what we love and we love what we know."

CONFLUENCE

3,220 CFS

The end is full of beauty. The Green comes to its confluence with the Colorado River in Canyonlands National Park, in what feels like the deep, guarded heart of canyon country. Three friends, Lauren, Robin, and Meghan, have come to finish the trip with me. We've switched to canoes so we can move faster through the receding flatwater. We have a week and 120 miles to paddle through Labyrinth and Stillwater Canyons to the Colorado. When we put on, at the edge of a browning golf course in Green River, Utah, the river has dropped again. It's slack and sand-colored in its banks, moving slowly.

The canyons are sunburn colored and streaked with dark desert varnish, sky high once we paddle out of the chippy, alkaline farmland around Green River. I am nostalgic already, clinging to the last few days on the river, even when they're long and hot. The bugs we'd been warned about come up in force, making even eating unpleasant some evenings, but still we float slowly, hunting clouds, trying to stretch the days. We take side hikes upstream, then jump in the river, drifting back to the boats, letting the current pull us down. My friends make it feel like vacation, they bring watercolors and cheese that isn't string, they weave sage crowns when we get stuck under a ledge in a thunderstorm, things I never would have done on my own. I am trying to hold onto every sunscratched second, my heart already sore from the idea of leaving the river behind, not sure if I'm claiming or claimed by the place.

Water is the only element that's public and private at the same time, environmental activist Linda Baker told me in the headwaters. Land doesn't move; no one owns air. Water exists in a tricky middle ground where everyone uses it, but no one is its keeper. "It's an interesting legal and moral concept," she said. "That's probably never going to be a question that will be resolved: Who owns the water? They say the water belongs to the state of Wyoming, but who is the state of Wyoming? It's us."

I think that sense of ownership is pivotal because water loss can feel like an empty threat when it's slow moving and far away. Many people are doing good in their own sphere, trying to make the best decisions they can see, but it's easy to abdicate big-picture responsibility because the burden is so dispersed. I think of Randy flood-irrigating to recharge the aquifer, or Kirk carefully managing Fontenelle. They're working within the history and responsibility they have, but the broader context is hard to see. I've felt that, too, my experience on the Green River butting up against what I've learned about the bigger Colorado River Basin. Both the good work and the bad news are true.

The river will be threatened and stretched further. That seems unavoidable. We've risked our rivers up to this point, so I don't see it stopping. Climate, unhelpful policy, and market forces will pull on it, degrade it, and thin it out. The embedded structures of water use, from flood irrigation to large-scale storage, are hard to break down, and we don't want to just shut them off. But there will be a point where the river can't stretch any more. Shortages will kick in; market forces will drive the reshuffling. Change in use has to be steady, but it also has to be serious enough that it motivates people to amend their behaviors. I'm worried, but I'm not hopeless, because I know that there are people trying to come to consensus at late-night meetings. There are ranchers saving fish, cities saving water. Things are changing, in small, hopeful ways.

It's bigger than this rugged, remote river. Loss of water is already affecting people across the U.S., especially in climatically and economically vulnerable places, like the tribal reservations of the Southwest, and drought is creating conflict and mass migrations of people around the globe. The future of water use here is a microcosm of a universal story, and I believe, even more strongly than I did when I first set out, that water shortages are going to be problematic worldwide. Access to water, at its core, is an equity issue in every sphere, from the environment to the economy. Some of the conflicts are unavoidable, and solutions will have to come from every stake: government, market change, and individual action.

The canyon is perfectly vertical in some places, eroded into spindly towers in others. At Bowknot Bend, a seven-mile-long gooseneck where the river doubles back to within a few hundred yards of itself, we stop and hike up to the narrow pass between the two limbs of the river. For years, paddlers would etch their names into the rock face at the top, claiming their place in the canyon. The Park Service doesn't let you deface the geology anymore, but you can still see the scratched-in record of river-running history. We scramble around looking for Hatches, Nevillses, and other names we recognize.

From up here I can see out to both braids of the river. The water is glowing green, deep and filled with light, bright against the ruddy rock. Everything feels vivid with the realization that I'm leaving soon. We hike down, eat lunch in a sliver of shade under a scorching rock, then paddle on around Bowknot into that ever-present wind.

After the bend, the canyon opens up. The rocks become green-gray and lavender in addition to red, and the channel gets wider. It's all sedimentary rock, sandstones and shales laid down by ancient oceans, slowly slipping away. There are striations in different strata, the layer cake of the shale and the scribbled sand of ancient seafloor encased in the crumbled-off boulders of Wingate Sandstone. We see more people as we get closer to the confluence: a group of Boy Scouts in canoes, a couple of kayakers.

We sleep on thin fingers of sandbars. A real storm—the kind where I'm counting the seconds between lightning and thunder—rolls through one night. It's the first precipitation I've seen in almost a month, and I worry more with the girls here than I did alone. I feel like I have to protect them, to preserve these last few days and make them perfect. In the morning, we wake up to a quickly eroding beach, much closer to the edge than we were the night before.

The risks are like this. They can be small and slow rolling, or they can break on you fast, hard to imagine until they're overhead. At the end, there had been no threat along the river, no event that left me fully fearful or shaken or doubting my ability to go on. It was a series of little things that kept me safe: micro-decisions, knowing when to ask for help, and trusting my gut when I thought I might be getting over my head.

I had set out, seven hundred miles ago, to see what happened to the river along its course. I'd received a lesson in the way that history shapes the stories we tell, and in the power of an unknown future. I couldn't quite pin down at the beginning what I wanted from the trip—in my head I had called it adventure—but as I get closer to the confluence I realize how much I wanted a stake, and something worth risking. I was disconnected from the future of the things I depended on. I felt scared and out of control.

At this point I've been through the hoop of the solstice: spring storms, the flush of high water, the blurred heat of desert high summer. The river is dropping. I can feel it tapering off, and getting colder, and I have more questions than I had going in. I had underestimated how much people cared, and how much work it takes to make a sustainable water system in a warming West without harming anyone. It is feasible, I think, but it's going to take upheaval, compassion, and a lot of communication from disparate water users.

I had to be gone, to be in it, to see the good and the bad. I learned that you can care about places, and want to protect them, in the abstract, but then you're fighting for abstractions. For me, it took that constant contact to start to understand their complexity. When you know what happens when side channels flood, filling the river with milky sediment, or what knuckle-winged blue herons look like when they take flight, you have a clearer sense of what's at stake. Randy, Albert, and the fly-fishing guides care as much as they do because they see the minutiae every day. It's muscle memory, woven into their neural pathways. They know what it would mean to lose the flows that sustain them. Now that I've lived on the Green, I understand it more. I've seen the things that change it: the dams, the inflows, the burble of agricultural runoff near Vernal, the washes that flood red when it storms. I've seen how even a few days in a canyon can change someone's perspective. It might not always take immersion, but I think it takes open eyes.

After so many miles, the confluence comes fast. On the last day, the river is turgid and thick orange after the previous day's storms. It carries runoff loaded so thickly with sediment that there's a layer of dirt and scum on top. We have a final lunch of pita and peanut butter under the ledges of a neighborhood of rock-chipped cliff dwellings, then push back out into the coppery swirl. The river takes a bend left, then a quick right. The canyon walls uplift and twist toward each other. "It looks like gates," Robin says, as our boats start to spin in the mingling currents.

We camp at the confluence and watch the Green, deep orange from the runoff, flow into the dun-colored Colorado. Their miscolored names are a tangible reminder that distant perception often shapes the way we think about the natural world, and that rivers are constantly changing within their bounds. It's one of the nicest camps I've had the whole trip, a wide alluvial fan, sandy, shady, mercifully bug free. We have chocolate, and a shook-up mini bottle of champagne, to celebrate before we pass out in a circle around the fire, sleeping bags slung next to each other as the embers burn out.

I wake up as soon as it's light and walk down to the point where the two rivers come together, tiptoeing in the dawn-damp sand, not ready to leave. Coming out is harder than going in. In the canyon I have a sense of place and a sense of purpose that comes from the rhythm of slowly moving downstream, and I don't feel ready to let that go.

I think water holds the question, "What do we want our future to look like?" and the answer is that "we," as a collective, doesn't exist. Trying to understand a whole river and predict the future was a faulty, self-centered goal. The adage about not stepping into the same river twice is true. The river changes, and you change. You think differently in the gravel-bedded headwaters than you do in the canyons of the confluence. As much as I'd like to think I saw the whole thing, I really got only a slice, temporal and thin. My trip was one series of stages at specific water levels. Out past the edge of the sandbar, the water is murky and fast, swirling with the Colorado's flow. It's a new river now, going out around the bend.

TIMELINE

1862   President Abraham Lincoln signs the Homestead Act

1869   John Wesley Powell takes his first trip down the Green and Colorado

1902   Congress passes the Reclamation Act

1908   *Winters v. United States* establishes the Winters Doctrine

1915   President Woodrow Wilson designates Dinosaur National Monument

1922   Seven basin states sign the Colorado River Compact

1928   The Boulder Canyon Act is passed, allocating the Lower Basin water

1934   President Franklin Roosevelt signs the Indian Reorganization Act

1938   Norm Nevills's first commercial river trip

1941   The Bureau of Reclamation proposes the Echo Park Dam

1944   Water Treaty with Mexico

1946   Bureau of Land Management is formed

1948   Upper Colorado River Basin Compact apportions the Upper Basin water

1948   Drilling starts in the Uinta Basin

1952   McCarran Amendment passes

1953   Bus Hatch receives first river concession permit in Dinosaur
        National Monument

1955   Secretary of the Interior Douglas McKay withdraws support
        for Echo Park Dam

1956   Central Utah Project authorized

1956   Colorado River Storage Project authorized by Congress

1962   Native fish poisoned with rotenone below Flaming Gorge

1963   Arizona v. California allocates water for tribes

1963   Fontenelle Dam built

1964   Flaming Gorge Dam completed

1968   Congress passes the Wild and Scenic Rivers Act

1973   Congress passes the Endangered Species Act

1976   The Federal Land Policy and Management Act designates
        public land management to the BLM

1988   Upper Colorado River Endangered Fish Recovery Program
        established

1992   Biological Opinion on operation of Flaming Gorge Dam
        issued

1992   Central Utah Project Completion Act enacted

1996   Ouray National Fish Hatchery established

2007   Secretary of the Interior Dirk Kempthorne signs Colorado
        River Interim Guidelines

2012   Colorado establishes recreational in-channel diversions

ACKNOWLEDGMENTS

Writing a book and setting off to do a long semi-solo river trip are both insular, self-centered undertakings, and ones that would be impossible without a watertight support network. I am so grateful for mine.

I am still amazed by how many people were deeply generous with their time and their ideas as I stumbled down the river, asking stupid questions. Eric Kuhn has been unflaggingly helpful in my understanding of water management. Mike Fiebig and Matt Rice at American Rivers, and Nathan Fey at American Whitewater, are some of the fiercest river advocates I know. Anne Castle is a role model for both her ability to broker consensus and to recite bawdy, epic poems around a campfire.

Randy Bolgiano, Albert Sommers, and Pat O'Toole changed my way of looking at the land, which might have been the most valuable lesson of this whole trip.

Tildon Jones, Heather Patno, and Tamara Naumann buck any kind of stereotype that government workers are cogs in a machine. They are all forces for understanding and connection. I'm glad to live in a world where Brad Udall, Jack Schmidt, Ted Kennedy, and Anya Metcalfe are all doing smart, thoughtful science.

Herm Hoops, Dennis Willis, and Stan Olmstead stand up for what they think is important in ways I find deeply inspiring. The folks at the Vernal O.A.R.S. outpost run a hell of a good show and didn't bat an eye at a weirdo solo writer.

Roy Webb is a wealth of knowledge about the history of the river. You should probably read his books before you read mine. Patty Limerick and Doug Kenney make me glad I went to grad school after all. So does Mike Koshmrl. Sorry if I ruined your relationship with Albert.

Zoe Sandler is the fairy godmother of agents, just cooler and with better taste. I never would have been brave enough to do this if she hadn't found me. Christie Henry believed in the idea and understood it when everyone else shirked an unpaddled river and a first-time author. Karen Merikangas Darling took the book on and turned it into something so much better, smoother, and smarter. Norma Sims Roche, who has a paddler's instincts along with copyediting chops, cleaned it up beautifully.

Nicole Grohoski, I'm grateful for your ability to make and navigate maps, even when they have unforeseen locks on them.

Will van der Veen, keeper of Friend Enforced Deadlines, most honest reader, dude who believed from the start. You're in my top five forever. Dutch too.

Meghan Barrier, Robin Bundi, Lauren Sancken, thank you for all the joy and all the accoutrements. MJ Carroll, how do you feel about . . .

My friends read drafts, fed me dinner when I was trying to make deadlines, sent me Twizzlers in weird corners of Utah, and otherwise kept me close to sane. Kate Silver-Heilman, Tom Weiss-Lehman, Kade Krichko, Hanady Kader, Nora Coughlan, Drea Templeton, the Littleriskys, Emily Anderson, Jess Henneman, Steph Pierce, Katie Lynch, Katie Cruickshank, I am so lucky.

My family is the best. Feel free to fact check that. Chris Hansman is a no-bullshit moral compass. Ellen Wernick provided support on every level, from fracking fact checks to water aerobics. My parents made me the kind of person who might want to do this in the first place, and held me together while I did. Mom, Dad, I don't even know where to start. Thank you, thank you, thank you.

NOTES

ON THE RIVER

1   U.S. Bureau of Reclamation, *Colorado River Basin Water Supply and Demand Study* (Washington, DC: Department of the Interior, 2012).
2   John Wesley Powell, *Report on the Lands of the Arid Region of the United States* (Washington, DC: Government Printing Office, 1872).

THE LAW OF THE RIVER

1   James Corbridge and Charles Wilkinson, "The Prior Appropriation System in Western Water Law: The Law Viewed through the Example of the Rio Grande Basin," Western Water Law in Transition (summer conference, Boulder, CO, June 3–5, 1985).
2   Colorado River Compact, 42 Statutes at Large (1921), p. 171.
3   Bills to Grant the Consent of the United States to the Upper Colorado River Basin Compact, H.R. 234, 81st Cong. (1948).
4   Eric Kuhn, "Joint West Slope Roundtables Risk Study and Results Summary" (Glenwood Springs: Colorado River District, 2016).
5   Governor Matthew Mead, *Wyoming Water Strategy* (Cheyenne: State of Wyoming, 2015).
6   Patrick Tyrrell, "Wyoming State Water Plan" (Cheyenne: Wyoming Water Development Office, 2002).
7   "Rural America, A Story Map," U.S. Census Bureau, accessed April 3, 2018, http://bit.ly/2hqcL40.
8   Wyoming Economic Analysis Division, "Regional Economic Profiles for U.S., Wyoming, and Counties: 1990 to 2016" (Cheyenne: Wyoming Economic Analysis Division, 2017).
9   Ranie Lynds and Rachel Toner, *Wyoming's Oil and Gas Resources* (Laramie: Wyoming State Geological Survey, 2015).

GROWING A CROP OF HUMANS IN THE DESERT

1   Act of May 20, 1862 (Homestead Act), Public Law no. 37–64, 12 STAT 392, https://www.loc.gov/rr/program/bib/ourdocs/Homestead.html.
2   Charles Dana Wilber, *The Great Valleys and Prairies of Nebraska* (Ann Arbor, MI: Daily Republican Print, 1881).
3   Family Farm Alliance, *Colorado River Basin Water Management— Principles and Recommendations* (Klamath Falls, OR: Family Farm Alliance, 2015).
4   Kate Greenberg, Lindsey Lusher Shute, and Chelsey Simpson, *Conservation Generation: How Young Farmers and Ranchers Are Essential to Tackling Water Scarcity in the Arid West* (Hudson, NY: National Young Farmers Coalition, 2016).
5   Spencer Blevins, "Valuing the Non-Agricultural Benefits of Flood Irrigation in the Upper Green River Basin" (master's thesis, University of Wyoming, 2015).

ALL THOSE PEOPLE HAVE TO EAT

1   U.S. Bureau of Reclamation, *Colorado River Basin Water Supply and Demand Study* (Washington, DC: Department of the Interior, 2012).
2   United Nations, *State of Food and Agriculture, 2016* (Rome, Italy: Food and Agriculture Organization, 2016).
3   Peter Gleick, *The World's Water*, vol. 8 (Washington, DC: Island Press, 2014).
4   U.S. Bureau of Reclamation, *Colorado River Basin Water Supply and Demand Study*.

THE ONLY WATERING HOLE IN THE WHOLE COUNTY

1   U.S. Census Bureau, *Growth in Urban Population Outpaces Rest of Nation* (Washington, DC: U.S. Census Bureau, 2012).

FLOWING UPHILL TO MONEY

1   Bradley Udall and Jonathan Overpeck, "The Twenty-First Century Colorado River Hot Drought and Implications for the Future," *Water Resources Research* 53 (2017): 2404–2418.
2   Southern Nevada Water Authority, "Conservation Facts and Accomplishments," May 22, 2018, https://www.snwa.com/drought-and-conservation/conservation-facts-and-achievements/index.html.
3   Utah Rivers Council, "Free Market Water," December 20, 2017, http://utahrivers.org/2015/09/02/free-market-water/.
4   Central Utah Completion Project Act, S.R. 1377, 106th Cong. (1999).

5   U.S. Census Bureau, *Utah Is Nation's Fastest-Growing State* (Washington, DC: U.S. Census Bureau, 2012).

6   Nathaniel J. Davis, *Order Denying Request for Rehearing* (Washington, DC: Federal Energy Regulatory Commission, 2012).

WHOSE RIGHTS?

1   D. Larry Anderson, *Utah's Perspective: The Colorado River* (Salt Lake City: Utah Division of Water Resources, 2002).

2   Colorado Water Conservation Board, *Colorado Water Plan* (Denver: Colorado Water Conservation Board, 2015).

CLAIMING AND RECLAMATION

1   World Commission on Dams, *Dams and Development: A Framework for Decision Making* (Berkeley, CA; World Commission on Dams, 2010).

2   Britt Alan Storey, *The Bureau of Reclamation: A Very Brief History* (Denver, CO; Bureau of Reclamation, 2016).

3   To Authorize the Secretary of the Interior to Amend the Definite Plan Report for the Seedskadee Project to Enable the Use of the Active Capacity of the Fontenelle Reservoir, H.R. 2273, 114th Cong. (2016).

4   Governor Matthew Mead, *Wyoming Water Strategy* (Cheyenne: State of Wyoming, 2015).

5   Karrie Lynn Pennington and Thomas V. Cech, *Introduction to Water Resources and Environmental Issues* (Cambridge: Cambridge University Press, 2010).

6   Mark Reisner, *Cadillac Desert: The American West and Its Disappearing Water* (New York: Viking, 1986).

7   Roy Webb, *Lost Canyons of the Green River* (Salt Lake City: University of Utah Press, 2012).

AFTER THE DAM

1   U.S. Bureau of Reclamation, *Final Biological Opinion on the Operation of Flaming Gorge Dam* (Denver, CO: U.S. Bureau of Reclamation, 1992).

2   U.S. Energy Information Administration, *State Energy Consumption Estimates 1960 Through 2015* (Washington, DC: U.S. DOE/EIA, 2015).

PROTECT THE GREEN RIVER AT ALL COST

1   Roy Webb, *Lost Canyons of the Green River* (Salt Lake City: University of Utah Press, 2012).

2   Mark Reisner, *Cadillac Desert: The American West and Its Disappearing Water* (New York: Viking, 1986).

3   American Society of Civil Engineers, *2017 Infrastructure Report Card* (Reston, VA: American Society of Civil Engineers, 2017).

4   U.S. Bureau of Reclamation, *Colorado River Basin Water Supply and Demand Study* (Washington, DC: Department of Interior, 2012).

5   Paul Grams and John Schmidt, "Geomorphology of the Green River in the Eastern Uinta Mountains, Dinosaur National Monument, Colorado and Utah," in *Varieties of Fluvial Form*, ed. Andrew J. Miller and Avijit Gupta (Hoboken, NJ: Wiley, 1999).

6   Theodore A. Kennedy, Jeffrey D. Muehlbauer, Charles B. Yackulic, David A. Lytle, Scott W. Miller, Kimberley L. Dibble, Eric W. Kortenhoeven, Anya N. Metcalf, and Colden V. Baxter, "Flow Management for Hydropower Extirpates Aquatic Insects, Undermining River Food Webs," *BioScience* 66, no. 7 (2016): 561–575.

THE MAP OF WHAT'S NEXT

1   Bradley Udall and Jonathan Overpeck, "The Twenty-First Century Colorado River Hot Drought and Implications for the Future," *Water Resources Research* 53 (2017): 2404–2418.

2   Dirk Kempthorne, *Interim Guidelines for the Operation of Lake Powell and Lake Mead* (Washington, DC: Department of the Interior, 2007).

3   American Society of Civil Engineers, *2017 Infrastructure Report Card* (Reston, VA: American Society of Civil Engineers, 2017).

LARVAL TRIGGERS

1   Anders Halvorson, *A Fully Synthetic Fish* (New Haven, CT: Yale University Press, 2010).

HUMANS ARE A SPECIES, TOO

1   Roy Webb, *Lost Canyons of the Green River* (Salt Lake City: University of Utah Press, 2012).

WHAT'S THE POINT OF A WILD RIVER?

1   Bruce Finley, "Colorado Shies from Big Fix as Proliferating People Seek More Water," *Denver Post*, July 9, 2015.

2   Wild and Scenic Rivers Act, Public Law 90-542, 82 Stat. 906 (1968).

3   Rebecca Manners, John Schmidt, and Michael Scott, "Mechanisms of Vegetation-Induced Channel Narrowing of an Unregulated Canyon

River: Results from a Natural Field-Scale Experiment," *Geomorphology* 211 (2014): 100–115.

THROUGH THE GATES

1 Roy Webb, *If We Had a Boat: Green River Explorers, Adventurers, and Runners* (Salt Lake City: University of Utah Press, 1997).
2 Roy Webb, *Riverman: The Story of Bus Hatch* (Salt Lake City: University of Utah Press, 2008).
3 Roy Webb, *If We Had a Boat.*
4 Logan Bockrath, director, *62 Years* (Thelonious Step, 2015).

WHAT IS IT WORTH?

1 Roy Webb, *If We Had a Boat: Green River Explorers, Adventurers, and Runners* (Salt Lake City: University of Utah Press, 1997).
2 Outdoor Industries Association, *The Outdoor Recreation Economy* (Boulder, CO: Outdoor Industries Association, 2017).
3 Tim James, Anthony Evans, Eva Madly, and Cary Kelly, *The Economic Importance of the Colorado River to the Basin* (Tempe: Arizona State University, 2012).
4 Concerning the Adjudication of Recreational In-Channel Diversions, CO SB06-037 (2006).
5 Utah Division of Water Resources, *Uintah Basin Plan* (Salt Lake City: Utah Division of Water Resources, 1999).

ENERGY AND POWER

1 Energy Policy Act 2005, Public Law No. 109-58, 2005.
2 Neil Kornze, *Recent Management of Oil and Gas Lease Sales by the Bureau of Land Management* (Washington, DC: Bureau of Land Management, 2016).
3 Utah Division of Oil and Gas, Department of Natural Resources, State of Utah, accessed December 19, 2017, https://oilgas.ogm.utah.gov/oilgasweb/.
4 U.S. Geological Survey Oil Shale Assessment Team, "Oil Shale Resources of the Eocene Green River Formation, Greater Green River Basin, Wyoming, Colorado, and Utah," U.S. Geological Survey Digital Data Series DDS-69-DD (2011).
5 Grant Willis, "The Utah Thrust System—an Overview," in *Geology of Northern Utah and Vicinity*, ed. L. W. Spangler and C. J. Allen, Utah Geological Association Publication 27 (Salt Lake City: Utah Geological Association, 1999), 1–9.

6   American Oil and Gas History Society, "First Utah Oil Well," updated 2017, https://aoghs.org/petroleum-pioneers/first-utah-oil-well/.

7   U.S. Energy Information Administration, "Wyoming State Energy Profile," last updated December 15, 2016, https://www.eia.gov/state/print.php?sid=WY.

8   U.S. Environmental Protection Agency, "Ensuring the Safe Management of Wastewater, Stormwater, and Other Wastes from Hydraulic Fracturing Activities," accessed December 19, 2017, https://www.epa.gov/hydraulicfracturing#providing.

9   Utah Division of Oil and Gas.

10   "Utah Oil and Gas Wells," Utah Department of Natural Resources, updated March 2017.

11   U.S. Geological Survey Oil Shale Assessment Team, "Oil Shale Resources."

12   K. A. Clark and D. S. Pasternack, "Hot Water Separation of Bitumen from Alberta Bituminous Sand," *Industrial and Engineering Chemistry* 24, no. 12 (1932): 1410-1416.

13   U.S. Government Accountability Office, "Water in the Energy Sector" (GAO 15-545, 2015).

WATER IS WHERE THE FIGHT IS

1   Utah American Indian Digital Archive, "History: The Northern Utes," University of Utah American History Center, updated 2008, https://utahindians.org/archives/ute/history.html.

2   Act of June 18, 1934, Public Law no. 73-383, 48 Stat. 984 (1934).

3   Winters v. United States, no. 158., 207 U.S. 564 (1908).

4   McCarran Amendment, 43 U.S.C. § 666 (1952).

5   Arizona v. California, 373 U.S. 546 (1963).

6   Robert Anderson, "Indian Water Rights and the Federal Trust Responsibility," *Natural Resources Journal* 46, no. 2 (2006): 399-437.

7   Central Utah Completion Project Act; S.R. 1377, 106th Cong. (1999).

8   Ute Indian Water Compact, Public Law 102-575, H.R. 429 (1992).

9   Central Utah Completion Project Act.

CLIMATE CHANGE IS WATER CHANGE

1   United States Global Change Research Program, *Global Climate Change Impacts in the United States* (Cambridge: Cambridge University Press, 2009).

2   U.S. Bureau of Reclamation, *SECURE Water Act Section 9503(c)—Reclamation Climate Change and Water 2016* (Washington, DC: U.S. Bureau of Reclamation, 2016).

3   Bradley Udall and Jonathan Overpeck, "The Twenty-First Century

Colorado River Hot Drought and Implications for the Future," *Water Resources Research* 53 (2017): 2404–2418.

4   U.S. Bureau of Reclamation, *Colorado River Basin Water Supply and Demand Study* (Washington, DC: Department of the Interior, 2012).

5   I. Stewart, D. Cayan, and M. Dettinger, "Changes in Snowmelt Runoff Timing in Western North America under a 'Business as Usual' Climate Change Scenario," *Climatic Change* 62 (2004): 217–232.

6   U.S. Bureau of Reclamation, *Colorado River Basin Water Supply and Demand Study*.

7   Monica Anderson, "How Americans View Environmental Issues," http://www.pewresearch.org/fact-tank/2017/04/20/for-earth-day-heres-how-americans-view-environmental-issues/ (April 2017).

THIS LAND IS YOUR LAND

1   Buzz Belknap and Loie Belknap Evan, *Belknap's Waterproof Desolation River Guide* (Evergreen, CO: Westwater Books, 2013).

2   Wallace Stegner, *Beyond the Hundredth Meridian* (New York: Penguin Random House, 1953).

3   Utah Public Lands Initiative Act, H.R. 5780, 114th Cong. (2016).

YOU CAN'T JUST SELL OUT TO A CITY

1   James Alton, *The River Knows Everything* (Salt Lake City: Utah State University Press, 2009).

2   Kate Greenberg, Lindsey Lusher Shute, and Chelsey Simpson, *Conservation Generation: How Young Farmers and Ranchers Are Essential to Tackling Water Scarcity in the Arid West* (Hudson, NY: National Young Farmers Coalition, 2016).

3   Spencer Blevins, "Valuing the Non-Agricultural Benefits of Flood Irrigation in the Upper Green River Basin" (master's thesis, University of Wyoming, 2015).

4   Buzz Belknap and Loie Belknap Evan, *Belknap's Waterproof Desolation River Guide* (Evergreen, CO: Westwater Books, 2013).

5   Ellen Hanak and Jelena Jezdimirovic, *California's Water Market* (Public Policy Institute of California, 2016), http://www.ppic.org/publication/californias-water-market/.

GETTING COMFORTABLE WITH RISK

1.   Michael Johnson, Lindsey Ratcliff, Rebecca Shively, and Leanne Weiss, *Looking Upstream: Analysis of Low Water Levels in Lake Powell and the Impacts on Water Supply, Hydropower, Recreation, and the Environment* (Boulder, CO: Western Water Policy Program, 2016).

INDEX